Australia's great mammal extinction debate rages on. Close to half the mammal extinctions across the planet over the last two hundred years have been of Australian species. The eighteen species lost include several beautiful wallabies and, of course, the thylacine.

The loss of these species is just the latest stage in a long decline that began in the last ice age. Huge wombats, possibly the largest-ever burrowing creatures, were once common, and there were marsupial 'lions' and a bird three times larger than the emu. Having debated the matter for a century and a half, experts still disagree on what caused the extinction of Australia's mammal megafauna.

Why did the thylacine and Tasmanian devil disappear from the mainland within the last few thousand years, why did the thylacine finally become extinct in Tasmania last century, and why have so many mammals been lost from inland Australia in the last two hundred years? These questions are hotly debated – and dingoes, fire, disease, sheep, rabbits, cats and foxes have all been blamed.

Chris Johnson reviews 50 000 years of Australia's environmental history in search of what really caused these waves of extinction. He has sifted and evaluated a multitude of studies across disciplines ranging from archaeology to ecology. His persuasively reasoned conclusion is that despite the great scale and complexity of the extinctions, a single process was responsible for them all.

Australia's Mammal Extinctions brings the great mammal extinction debate to readers with no specialist knowledge of the many fields it covers. The splendid illustrations capture the poignancy of what we have lost, and the clearly argued, beautifully written text challenges us to use our knowledge to conserve those mammals that survive.

Chris Johnson is a Professor in the School of Tropical Biology at James Cook University, where he teaches and researches ecology and wildlife conservation.

Australia's Mammal Extinctions

A 50 000 year history

Chris Johnson

CAMBRIDGE UNIVERSITY PRESS
Cambridge, New York, Melbourne, Madrid, Cape Town, Singapore, São Paulo

Cambridge University Press
477 Williamstown Road, Port Melbourne, VIC 3207, Australia

Published in the United States of America by Cambridge University Press, New York

www.cambridge.org
Information on this title: www.cambridge.org/9780521686600

First published 2006

Printed in China by Bookbuilders

Typeface Bembo 11/14 point System QuarkXPress®

A catalogue record for this publication is available from the British Library

National Library of Australia Cataloguing in Publication data
Johnson, Chris, 1958–
Australia's Mammal Extinctions: A 50 000 Year History.
ISBN-13 978-0-521-84918-0 hardback
ISBN-10 0-521-84918-7 hardback
ISBN-13 978-0-521-68660-0 paperback
ISBN-10 0-521-68660-1 paperback
1. Extinct mammals – Australia. 2. Extinction (Biology) – Australia I. Title.
569.0994

ISBN-13 978-0-521-84918-0 hardback
ISBN-10 0-521-84918-7 hardback

ISBN-13 978-0-521-68660-0 paperback
ISBN-10 0-521-68660-1 paperback

CONTENTS

PREFACE

AUSTRALIANS ARE OFTEN told that for the last two hundred years their country has had the world's worst record of mammal extinctions. Unfortunately this is true. What is less well known is that the recent disappearances of species like the desert rat-kangaroo, pig-footed bandicoot and white-footed tree-rat are part of a very long process of decline of the continent's mammal fauna that began about 50 000 years ago. Understanding the cause (or causes) of this long decline is a profoundly important problem. It is at the centre of some of Australia's most vexed environmental debates, such as whether ancient Aboriginal people hunted giant mammals to extinction, and how those first people changed Australian ecosystems by their use of fire. We also need to understand what caused the more recent declines, because as well as resulting in many extinctions they left in their wake a large collection of rare and threatened mammals. Much conservation effort in Australia is devoted to saving mammal species on the brink of extinction. If we want to succeed in this, we need to understand what pushed them to the brink in the first place.

There has been a lot written about mammal extinction and conservation in Australia, but this literature is widely scattered. The debate over prehistoric extinctions has mostly been conducted in the pages of journals of archaeology and palaeontology, although it has occasionally broken out into popular books and the mass media. People who study recent extinctions and the management of threatened mammal species usually publish their work in journals of ecology or conservation, or in specialised scientific volumes. There is a lot of this and it is of high quality, but very little of it refers to the prehistoric antecedents of the recent declines and extinctions. In fact, with the exception of Tim Flannery's *The Future Eaters*, the ancient and recent extinctions of Australian mammals have been treated as completely separate and independent problems.

My aim in this book is to bring these fields of prehistoric and historic mammal extinctions together, and to ask if they fit within a common framework of explanation. In other words, I want to find out if they are part of the same story, and to recount the events of that story in their proper order and relationship to one another. At its core, this effort is based on a thorough investigation of all the primary evidence. I have delved deeply into the literature on Australian palaeontology, archaeology, climate and vegetation change, and contemporary ecology, and that information is used to produce a complete history of Australian mammals focusing on that critical period of the last 50 000 years.

I hope *Australia's Mammal Extinctions* will make sense as a history of Australia: it is the history of Australia as registered in changes to the continent's mammal fauna. As well as putting together this historical narrative, I use the primary evidence to test competing hypotheses on what caused so many mammals to go extinct. And at the end of the book I suggest a version of events that links all of the extinctions together.

The book covers some quite technical ground, but I have tried to write it in a way that will be easily understood by readers who have no specialist knowledge in any of the disciplines mentioned above. I have avoided technical language as far as possible, and for those jargon terms that survived and might cause difficulties there are brief explanations in a short glossary.

Writing this book has been a big job, but I have been helped by many people. In particular I want to thank, for their generosity in reading and commenting on various chapter drafts, David Bowman, Barry Brook, Richard Gillespie, Peter Jarman, Paul Martin, Robyn Montgomery and Stephen Wroe; I am especially grateful to Peter who read about half the manuscript and to Barry and Robyn who read the whole thing. The book was improved a great deal through discussions with all of these people, and also with Ian Abbott, Chris Dickman, Will Edwards, Judith Field, Bill Foley, Richard Fullagar, Betsy Jackes, Menna Jones, the late Jill Landsberg, Peter Latz, Jon Luly, Ben Moore, Anne Musser, Des Nelson, Bob Paddle, Matt Pye, Gavin Prideaux, Andy Purvis, Bert Roberts, Grahame Walsh, Michelle Waycott and Rod Wells.

Several people generously provided data sets, allowed me to cite their unpublished data, or sent unpublished manuscripts: for such favours I thank Matthew Cupper, Joe Dortch, Tim Flannery, Scott Hocknull, Jon Luly, Marie Murphy, Bob Paddle, Gavin Prideaux, Andrew Rowett, Jeff Short, Michelle Waycott and Rod Wells. Peter Murray is not only responsible for much of the primary palaeontological work that made it possible to write a book like this, but he also provided some of his stunningly beautiful drawings of extinct species, while another extraordinary palaeontologist/artist, Anne Musser, contributed some of her equally beautiful paintings. Thanks to you both, and to Ashley Field for his miniatures of living marsupials. Some original and very special photographs were provided by Michael Cermak, George Chaloupka, Ken Johnson, Jen Martin, Euan Ritchie and Karl Vernes.

I began thinking seriously about this project during a six-month sabbatical from James Cook University. I am grateful to JCU for providing that opportunity, as I am to the University of Sydney, and Chris Dickman in particular, for hosting my sabbatical visit. My research has been supported by a number of organisations, but especially James Cook University and the Australian Research Council. At Cambridge University Press, Kim Armitage, Jean Dunn and Kate Indigo thoughtfully and expertly guided the manuscript through editing and production – thank you.

Finally, I want to thank most sincerely, for their patience and support in many ways, Henry, Lachlan, Philippa and, most of all, Robyn.

CHRIS JOHNSON
Townsville

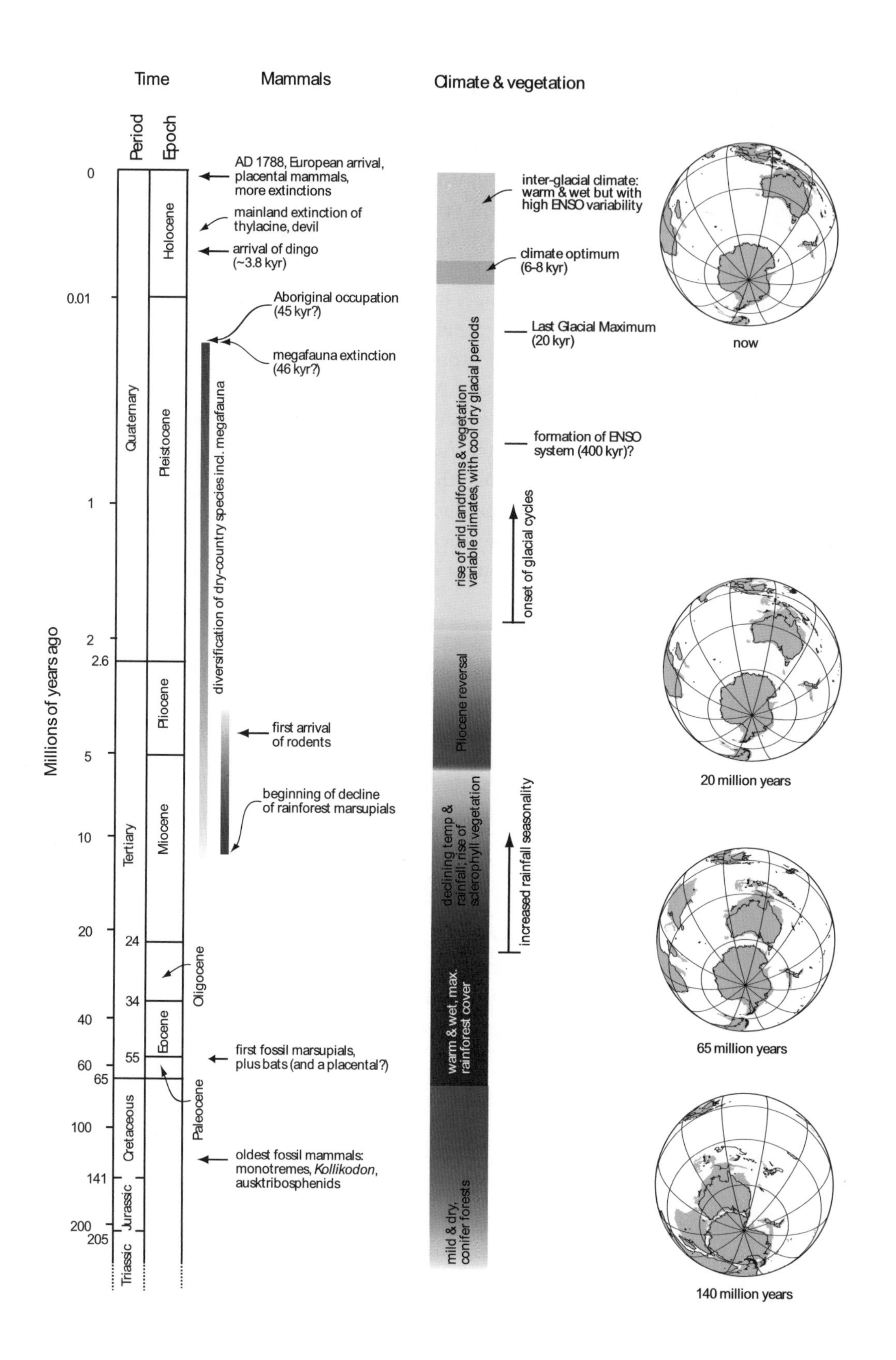

Time
Mammals
Climate & vegetation
Period
Epoch
Millions of years ago
0
0.01
1
2
2.6
5
10
20
24
34
40
55
60
65
100
141
200
205
Quaternary
Tertiary
Cretaceous
Jurassic
Triassic
Holocene
Pleistocene
Pliocene
Miocene
Oligocene
Eocene
Paleocene
AD 1788, European arrival, placental mammals, more extinctions
mainland extinction of thylacine, devil
arrival of dingo (~3.8 kyr)
Aboriginal occupation (45 kyr?)
megafauna extinction (46 kyr?)
diversification of dry-country species incl. megafauna
first arrival of rodents
beginning of decline of rainforest marsupials
first fossil marsupials, plus bats (and a placental?)
oldest fossil mammals: monotremes, Kollikodon, ausktribosphenids
inter-glacial climate: warm & wet but with high ENSO variability
climate optimum (6-8 kyr)
Last Glacial Maximum (20 kyr)
formation of ENSO system (400 kyr)?
rise of arid landforms & vegetation variable climates, with cool dry glacial periods
onset of glacial cycles
Pliocene reversal
declining temp & rainfall; rise of sclerophyll vegetation
increased rainfall seasonality
warm & wet, max. rainforest cover
mild & dry, conifer forests
now
20 million years
65 million years
140 million years

1

A brief history of Australia's mammals

OOLACUNTA

In December 1931 the mammalogist Hedley Herbert Finlayson travelled from Adelaide to Appamunna on the northwestern edge of the Simpson Desert in search of the desert rat-kangaroo *Caloprymnus campestris*. This species was something of a mystery to science. It had been described in 1843 on the basis of two specimens sent to John Gould in London from an unspecified location, but had not been recorded again until L. Reese, the owner of Appamunna Station, sent Finlayson a specimen and reported some sightings of the animal, which was known to the local Aborigines as 'oolacunta' (Figure 1.1). Finlayson wanted to learn more about oolacunta and to collect a series of specimens for the South Australian Museum. With Reese's help he recruited four knowledgeable Aboriginal men as expedition members and guides, and followed them into what they said was prime oolacunta country. This turned out to be perhaps the most extreme environment in an extraordinarily harsh region: a chain of gibber plains fringed by low sandhills, with no water, very little plant cover, and practically no shade (Plate 1).

Finlayson's party decided they would search for oolacunta on horseback, beating across country and following tracks if any could be found among the stones. If an oolacunta broke from cover it would be pursued, one rider chasing it at full gallop and the others replacing the lead rider when his horse began to tire. This plan worked, and the group soon put up their first animal. Finlayson (1935a) provides a vivid account of this first sighting:

> it was only after much straining of the eyes that the oolacunta could be distinguished – a mere speck, thirty or forty yards ahead. At that distance it seemed scarcely to touch the ground; it almost floated ahead in an eerie, effortless way that made the thundering horse behind seem, by compari-

OPPOSITE Summary of major events in the history of Australian mammals, in relation to the geological time scale, changes in climate and vegetation (shown on the vertical bar, where darker shading indicates more warm and wet conditions), and the changing location of Australia in the southern hemisphere.
Notes: Information on climate and vegetation from McGowran *et al.* (2000) and Kershaw *et al.* (2000). The date of 2.6 million years ago for the beginning of the Pleistocene follows Kershaw *et al.* (2000).

FIGURE 1.1 The desert rat-kangaroo, or oolacunta, photographed by Finlayson (1932).

son, like a coal hulk wallowing in a heavy sea. They were great moments as it came nearer; moments filled with curiosity and excitement, but with a steady undercurrent of relief and satisfaction. It was here!

This little animal led them on an astonishing chase: they did not catch it until it had run twelve miles and tired out three horses, whereupon it died of exhaustion. The party collected seven specimens in this way, and all showed the same amazing tenacity. Another two animals, a mother and young, were caught by hand from their nest by one of the Aborigines who had hunted oolacunta before and could remember when it had been abundant. In all, seventeen animals were sighted in a week of searching (Finlayson 1932), a tally suggesting that oolacunta was locally common. Further specimens were received from the area until 1936. Reliable sightings continued until the mid-1950s, with some possible sightings as late as the 1970s, but there have been no more confirmed records and oolacunta is now thought to be extinct (Carr & Robinson 1997). What could have caused an animal like this – a survivor in one of the harshest environments in Australia, with the heart for such extraordinary feats of endurance – to disappear for ever?

THE EXTINCTION PROBLEM

The loss of oolacunta was just one of the latest in a long series of extinctions that have depleted Australia's mammal diversity in the geologically recent past. One hundred thousand years ago there were at least 340 species of land mammals in Australia; 67 of them are now extinct. These extinctions came in three waves. First, some time in the last glacial cycle (the late Pleistocene, between 130 000 and 10 000 years ago), over 50 species of mainly very large marsupials disappeared. Next, in the Holocene (between 10 000 and 200 years ago) the remaining large carnivores declined to extinction on the mainland. Finally, in the two hundred years or so since European settlement a further ten marsupial species and eight rodents have gone extinct, and almost one in four marsupial species is now threatened with extinction (Maxwell *et al.* 1996).

Mammal faunas on other continents have suffered recent extinctions, but not as badly as this. For example, there were megafauna extinctions in the late Pleistocene in South America (and in many other parts of the world) just as there were in Australia, but the South American loss was only about 10 per cent of the mammal fauna compared with 18 per cent in Australia. There was no parallel in South America to the Holocene decline of Australian large carnivores, and no South American mammals have gone extinct since the arrival on that continent of Europeans. Of the 40 mammal species worldwide known to have vanished in the last two hundred years, almost half have been Australian (MacPhee & Flemming 1999).

What is it about Australian mammals, or the Australian environment, or the nature of human impact on them, that can explain these high extinction rates? This book reviews the history of environmental change in Australia in a search for the answer to this question. Its three main sections deal with the declines of the late Pleistocene, Holocene and European periods respectively. Nonetheless, I am concerned to see this history as a whole: I will argue that the three waves of extinction were part of a single unfolding sequence of cause and effect, and that to understand the high rate of extinction and endangerment of Australian mammals in recent history we must begin by looking back at the events of the late Pleistocene and Holocene.

ORIGINS AND EVOLUTION

Australia has a unique and special mammal fauna. It is only here (along with New Guinea) that all three of the world's major lineages of living mammals – the monotremes, marsupials and placental mammals – can be found together. The monotremes occur nowhere else, and the Australo-Papuan region is the global centre of marsupial diversity. The long period of evolution of these groups in Australia and New Guinea produced a spectacular diversity of forms paralleling the great radiations of placental mammals over the rest of the globe. Many Australian mammals independently evolved remarkable similarities to their placental counterparts: gliding possums are strikingly similar to flying squirrels, for example, and marsupial moles look almost identical to the golden moles of Africa. Others evolved to gigantic size, producing rhinoceros-sized herbivores and lion-sized predators. On the other hand, some Australian mammals are unlike any species found elsewhere in the world. There is no other mammal quite like a wombat (a large herbivore that lives in a burrow), a Tasmanian devil (a small bone-crushing scavenger) or a honey possum (a specialised tube-nosed non-flying nectar feeder), not to mention outstanding oddities such as the platypus.

MYSTERY MAMMALS OF THE CRETACEOUS

The earliest fossil mammals from Australia date from the middle Cretaceous, around 105 to 120 million years ago. The small collection of fossils from this period consists of two monotremes plus several species that are not closely related to any living mammals. One of the monotremes, *Steropodon galmani* from Lightning Ridge, was probably platypus-like but was larger than the living platypus and may have weighed close to three kilograms; the other, *Teinolophos trusleri* from Flat Rock in Victoria, was a tiny relative of *Steropodon* with an estimated total length of only nine centimetres (Long *et al.* 2002). The enigmatic *Kollikodon ritchiei*, also from Lightning Ridge, was originally described as a monotreme (Flannery *et al.* 1995) but a more detailed analysis shows that it may be an entirely new type of mammal only distantly related to monotremes (Musser 2003, and personal communication). It was about the same size as *Steropodon*, making these two species among the largest mammals of their time anywhere in the world.

The other two mammals known from the Cretaceous of Australia belong to a group known as the ausktribosphenids, so far known only from the Flat Rock fossil site. These were small shrew-like creatures less than ten centimetres in total length. Just how they are related to the other mammal groups is a great puzzle. Rich *et al.* (1997) argued that they were placental mammals, a claim that challenges the accepted view of the history of the placental mammals, which are otherwise known only from the northern hemisphere until just after 65 million years ago (Rich *et al.* 1999). The controversy ignited by Rich *et al.*'s interpretation of the ausktribosphenids has not yet died down and many alternative affinities have been proposed for them, including that they were actually monotremes.

The various interpretations are summarised by Woodburne (2003), who thinks the ausktribosphenids belong in an ancient and now completely extinct lineage of eutherian mammals allied to the living placentals.

We still know very little about Australia's Cretaceous mammals, but the diversity represented by the five species described so far, which have been collected from only two fossil sites, suggests a rich fauna with a long evolutionary history reaching back probably into the Jurassic. All that remains are the living monotremes, the platypus and the echidnas. Monotremes are great survivors, being the last representatives of the ancient mammal group Prototheria (as distinct from the Theria, the group to which all other living mammals belong). The platypus lineage, which connects *Steropodon* and the living *Ornithorhynchus*, has been splashing about in freshwater streams in Australia for at least 120 million years.

From 105 to 55 million years ago no fossil Australian mammals are known. Presumably, the ausktribosphenids died out during this period. True platypuses (family Ornithorhynchidae) first appeared in the Australian fossil record about 25 million years ago, and echidnas somewhat later. The echidnas became moderately diverse, and before the late Pleistocene extinctions there were at least six species spread through Australia and New Guinea (Musser 1999).

The ancient monotremes and ausktribosphenids lived on a continent that for the most part was warmer than now and had a more uniform climate, except that the far south, which was still in contact with Antarctica, was cold and seasonal. The land was covered by forests of tall conifers from the families Araucariaceae and Podocarpaceae. These forests were evidently quite open-canopied and in most places the diversity of understorey vegetation was high (Henderson *et al.* 2000). From the late Cretaceous the climate appears to have become warmer and wetter, and there was an expansion of dense rainforests dominated by flowering plants. By the Eocene, about 50 million years ago, wet rainforest covered most of the continent. Species of *Nothofagus* (subgenus *Brassospora*) were typical of these Eocene forests, but they now occur only in New Guinea and New Caledonia in very warm, continuously wet environments. Another indicator of the Eocene climate of Australia was the very wide distribution of *Gymnostoma*, a rainforest-restricted lineage of the Casuarinaceae that now occurs only in a small area on the Atherton Tableland in the wet tropics of north Queensland (Hill 1994). *Nothofagus* and *Gymnostoma* grew even in the now-arid centre of the continent (Kershaw *et al.* 1994). It was into these vast, dense, wet and warm rain-forests that the marsupials wandered when they completed their journey to Australia.

THE LONGEST JOURNEY: THE MARSUPIALS

The history of the marsupials can be traced back to a 125 million-year-old fossil from northeastern China, *Sinodelphys szalayi* (Luo *et al.* 2003). *Sinodelphys* was a small, agile, insectivore/carnivore that was well-adapted for tree climbing. While not strictly a marsupial, it is the earliest known member of the Metatheria, the major mammal group of which the marsupials are the living representatives. Metatherians appeared in western North America about 115 million years ago, probably having crossed from eastern Eurasia via a land-bridge that opened between Alaska and Siberia in the mid-Cretaceous (see the PALEOMAP project web site: www.scotese.com). These early mammal fossils from western North America include *Kokopellia juddi*, the earliest mammal actually classified (by some authorities) as a marsupial (Cifelli & Muizon 1997).

From 115 to 65 million years ago marsupials flourished in North America along with placental mammals (Cifelli & Davis 2003). They entered South America just after the end of the Cretaceous, between 65 and 60 million years ago (Muizon *et al.* 1997). At that time there was no land connection between North and South America, but a chain of volcanic islands in what was probably a shallow ocean separating the two continents may have allowed mammals to island-hop from north to south, presumably by accidental rafting on flotsam (Lillegraven 1974). South America was in contact with Antarctica, which in turn was in contact with Australia, providing the marsupials with a southern route to Australia. Fossil marsupials have been found in Antarctica, at Seymour Island on the Antarctic Peninsula near South America (Goin *et al.* 1999). These fossils are thought to be about 45 million years old, but they resemble taxa from much earlier South American faunas, except for two that are distinct from any known South American species and probably diverged after entering Antarctica. On these grounds Goin *et al.* (1999) suggest that the original members of this fauna may have colonised Antarctica about 60 million years ago.

The oldest fossil marsupials in Australia are from a site at Tingamurra near Murgon in southeastern Queensland that is dated to at least 55 million years ago (Archer *et al.* 1993; Godthelp *et al.* 1992). Most of the Tingamurra marsupials are unlike South American or Antarctic forms, suggesting that they underwent a significant period of evolution after reaching Australia, and therefore that they arrived substantially earlier than 55 million years ago. So, the evidence from both the Antarctic and Australian fossil record is that marsupials spread through Antarctica and into Australia

PLATE 1 Habitat formerly occupied by oolacunta, the desert rat-kangaroo *Caloprymnus campestris*, near Birdsville in southwestern Queensland. The animal primarily inhabited the gibber plains in the background. (photograph by Jon Luly)

PLATE 2 Long-tailed planigale *Planigale ingrami*, a tiny insectivorous marsupial. The diversity of such small insectivores and other dasyuromorph marsupials exploded at the end of the Miocene as dry, open habitats expanded in Australia. (photograph by Euan Ritchie)

PLATE 3 Golden-backed tree-rat *Mesembriomys macrurus*, a large arboreal and somewhat squirrel-like rodent from northern Australia. (photograph by Euan Ritchie)

PLATE 4 Queensland pebble-mound mouse *Pseudomys patrius*, a tiny rodent that builds large mounds of pebbles over its burrows. (photograph by Michael Cermak)

very soon after they first entered South America (Woodburne & Case 1996).

In most cases it is not possible to trace the connections of particular lineages of marsupials across South America, Antarctica and Australia. The one exception is the family Microbiotheriidae, which has only one living species, *Dromiciops australis* or the monito del monte (little monkey of the mountains) of southern South America. The Seymour Island fauna includes one microbiotheriid and there are two more species in the Tingamurra fauna (Long *et al.* 2002). The living *Dromiciops* is restricted to *Nothofagus* forest, which was the dominant vegetation in Antarctica at the time the Seymour Island fossil beds were laid down and in eastern Australia at the time of the Tingamurra fauna. Although the microbiotheriids are now extinct in Australia, the living *Dromiciops* is more closely related to Australian than to South American marsupials (Cardillo *et al.* 2004). Probably the majority of living Australian marsupials are descended from a common ancestor with microbiotheriids.

Placental mammals also crossed from North to South America just after the end of the Cretaceous. The oldest known placental mammals in South America were rather less diverse and less divergent from their North American ancestors than were the South American marsupials of the time, suggesting that they crossed into South America later than the marsupials (Muizon *et al.* 1997). There are fossil placentals along with marsupials in Antarctica, but again their low diversity and close affinities with South American species suggest that they crossed into Antartica after the marsupials (Goin *et al.* 1999). Finally, the Tingamurra fauna of Australia includes a species, *Tingamurra porterorum*, that was described as a placental mammal by Godthelp *et al.* (1992). Apart from early bats (see below) this is the only placental identified in the Tertiary of Australia until the arrival, much later, of rodents. However, the interpretation of *T. porterorum* as a placental mammal rather than a marsupial has been challenged (Szalay 1994; Woodburne & Case 1996). Godthelp's announcement of a placental mammal in the company of the early marsupials of Australia was an event deeply felt by Australian mammalogists, because it seemed to refute the old opinion that marsupials had been successful in Australia only because of the absence of competition from (superior?) placental mammals. Given the uncertainty over what *Tingamurra* was, however, it remains very possible that placental mammals did not complete the journey along the southern route into Australia in the early Tertiary.

Why might marsupials but not placental mammals have managed to cross into Australia from Antarctica? There could be two explanations.

First, the ocean gap between the two continents may have begun to form as early as 70 million years ago (Smith *et al.* 1994), and the likelihood that a terrestrial mammal from Antarctica could find its way to Australia would have diminished steadily after that time. It seems that the marsupials got into South America before placental mammals, and also beat the placentals into Antarctica. Maybe they reached eastern Antarctica in time to make the crossing into Australia, but the placentals did not. Second, it is possible that even if marsupials and placentals reached eastern Antarctica at the same time, the marsupials were better fitted to make the journey north if that required a sea crossing. The marsupials known from Antarctica were all small, probably less than 350 grams, and they were arboreal insectivores or frugivores (Goin *et al.* 1999). Probably they were more likely to be cast out to sea on flotsam than the placentals, which were larger herbivores; such rafts might also have carried insects, and the marsupials could have stayed alive by eating them as they drifted north.

As far as we know the first marsupials to reach Australia encountered no terrestrial mammals other than a few monotremes, which were already quite specialised creatures. Their subsequent evolutionary history is a classic example of diversification to exploit the many opportunities offered by an unoccupied new land mass. Because the marsupials arrived in Australia when rainforests were expanding, their initial diversification was predominantly of forms adapted to complex forest. Unfortunately we know nothing about the early development of this rainforest fauna, because after Tingamurra there is a 30 million-year gap in the fossil record of Australian mammals (Archer *et al.* 1999). The mid-Oligocene faunas that emerged from this palaeontological dark age were remarkably rich in species. Marsupial diversity continued to increase until it reached an all-time high in the Early Miocene, when there were about 50 per cent more families in existence than in the last few million years (Archer *et al.* 1999)

At that point the climate began to turn cooler, drier and more variable. Eventually this trend produced fully arid conditions in much of inland Australia (McGowran *et al.* 2000; Quilty 1994). The major cause of this change was a re-arrangement of ocean circulation in the southern hemisphere following the final breaking apart of Gondwana. Before the break-up, cold currents flowing along the coast of Antarctica were turned northward when they struck the west coasts of South America and Australia. Their waters then circulated through the tropics before returning south, transporting tropical warmth to the Antarctic region. As Australia and South America moved away from Antarctica, a closed circum-Antarctic current formed. The waters of the southern ocean were therefore separated from those of the tropics, and the result was a drop in temperature of the southern ocean, the freezing over of Antarctica, and the establishment of cooler conditions in southern Australia. Latitudinal gradients in temperature became steeper, and climate zones became more strongly differentiated. At the same time the northward drift of Australia removed most of its central and northern parts from the influence of moisture-laden westerly winds and exposed them to drier and warmer subtropical high pressure systems (Bowler 1982; Bowman 2000).

With the drying and cooling of the climate the Eocene rainforests retreated and were replaced by more open and drier forests and woodlands, and in some places by shrublands and grasslands (Figure 1.2). These dry forests and woodlands were often dominated by Casuarinaceae (casuarinas) and they included *Acacia*, Asteraceae (daisies) and Chenopodaceae (chenopods, i.e. bluebush and saltbush) as indicators of the expansion of a shrub and herb layer under an opening forest canopy. By the end of the Pliocene the vegetation of Australia broadly resembled that of today in its composition and in the geographic distribution of structural types. One major difference was that rainforests were still rather more widespread and, in particular, there was a more extensive development of dry rainforest dominated by conifers, especially Araucariaceae. Another was that in the dry forests Casuarinaceae were often more abundant than eucalypts.

This drying trend caused the extinction of some specialised rainforest-dependent marsupial groups, and the family-level diversity of marsupials declined from the mid- to late-Miocene, never to fully recover. But the new environmental conditions also stimulated the diversification of dry-country species. In the dasyuromorph (insectivorous and carnivorous) marsupials, for example, there was a surge of diversification in the mid–late Miocene that coincided with the expansion of dry vegetation across inland and northern Australia (Figure 1.2). Most of the species produced in this late Miocene burst of evolution were small insectivores of grasslands, shrublands and open woodlands: animals such as dunnarts, planigales, and so on (see Plate 2).

The drier environments of the Pliocene and Pleistocene also provided many opportunities for large terrestrial herbivores, among which the Macropodidae (kangaroos) were the most successful. Macropodid diversity exploded and by 100 kyr ago there were at least 95 species of kangaroos and wallabies in Australia and New Guinea (Cooke & Kear 1999; Prideaux 2004). The main evolutionary trend in this diversification was increased body size and the accumulation of adaptations – in the teeth, skull and digestive system – for feeding on tough, abrasive plant material of low nutritional value. This was carried furthest in the sub-family

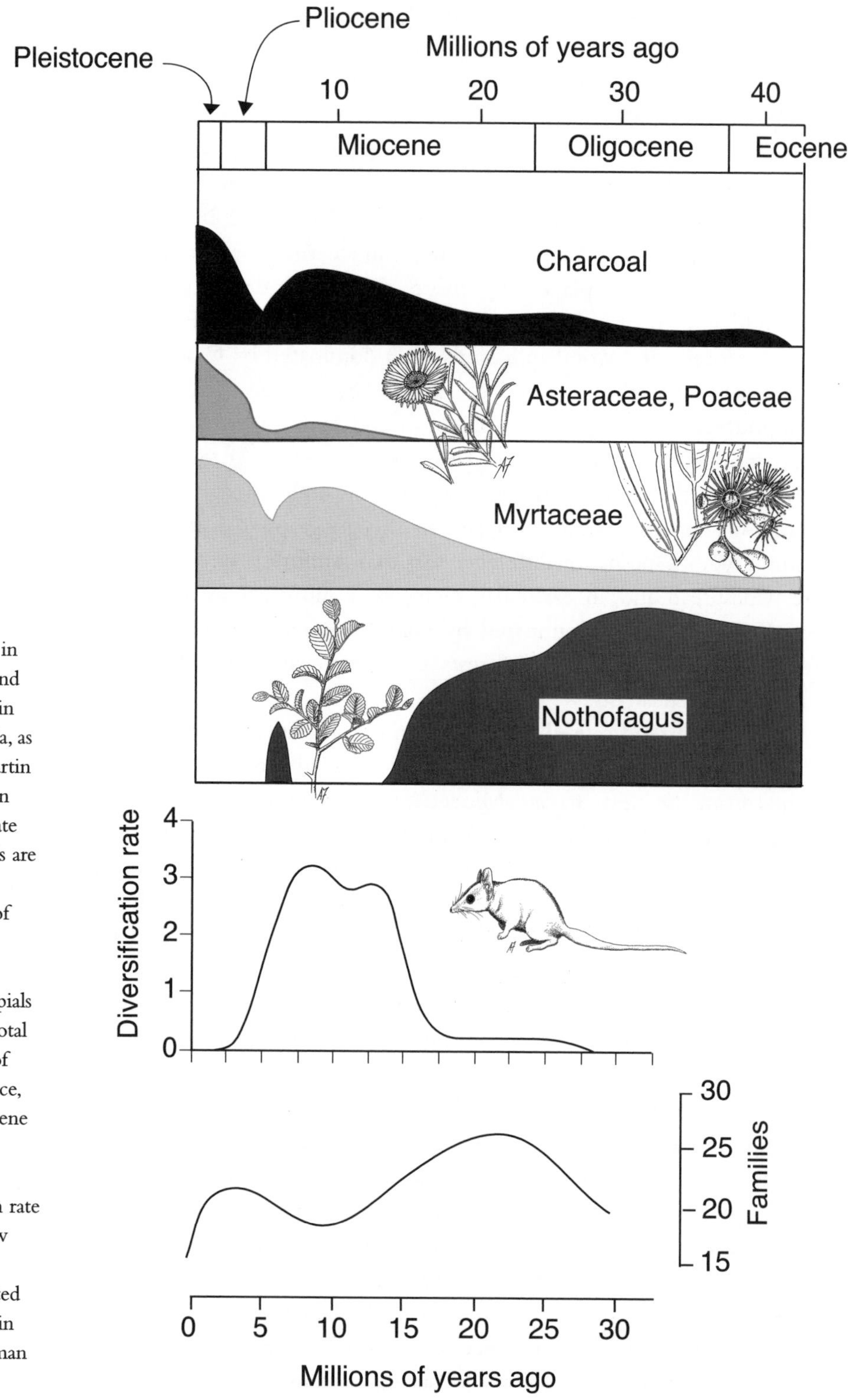

FIGURE 1.2 Changes in vegetation, climate and charcoal abundance in southeastern Australia, as reconstructed by Martin (1990) for the eastern Murray Basin. Climate and vegetation trends are compared with diversification rates of dasyuromorph (carnivorous and insectivorous) marsupials and changes in the total number of families of marsupials in existence, from the late Oligocene to the present (from Archer *et al.* 1999). *Notes:* Diversification rate is the number of new lineages arising per million years, estimated from the phylogeny in Krajewski & Westerman (2003).

Sthenurinae, a group of large kangaroos with many species in shrublands and dry woodlands. Other families of large herbivores, notably the Vombatidae (wombats) and Diprotodontidae (diprotodons), also increased in body size and species richness, and there was a correlated evolution of large predators in the families Thylacinidae (marsupial 'wolves') and Thylacoleonidae (marsupial 'lions'). These large herbivores, and the large carnivores that preyed on them, constitute the bulk of the 'marsupial megafauna' that is described in more detail in the next chapter.

Around two and a half million years ago – the beginning of the Pleistocene – global temperatures began to fluctuate in regular cycles associated with advances and retreats of northern hemisphere ice sheets. In Australia these glacial cycles produced alternations between cool and dry conditions in glacial periods, and warm and wet conditions in interglacials. These cycles had profound effects on the distribution of vegetation, which are described in detail in Chapter 5. They also drove sea levels up and down as vast quantities of sea water were frozen in the northern hemisphere ice sheets during glacial periods and then released as the ice melted during interglacials. Reconstructions of sea level around the north of Australia through the last glacial cycle show that the sea was at its present stand, or slightly higher, for a brief period around 120 kyr ago; sea levels then trended gently downward until at 30 kyr ago there was a rapid fall from 70 metres to around 130 metres below present-day levels. The sea began rising again just after 20 kyr ago and stabilised about 7 kyr ago (Lambeck *et al.* 2002).

Because of falling sea levels Australia became a much bigger place under glacial climates. Much of the continental shelf was exposed as dry land; in the north the Carpentarian Plain emerged and connected Australia to New Guinea, and in the south the Bassian Plain bridged the mainland and Tasmania. This single enlarged landmass is known as Greater Australia (Figure 1.3). For most of the Pleistocene the Australia/New Guinea region bore a much closer resemblance to the map of Greater Australia shown in Figure 1.3 than to the present-day map. This is an important point, because critical events like the extinction of the megafauna and the arrival of people were played out over the land surface of Greater Australia rather than on the distinct land masses of New Guinea, Australia and Tasmania.

During the early Tertiary most of what is now New Guinea was under water, but an episode of uplift in the Oligocene exposed a large area of southern and western New Guinea, which was connected to northern Australia and which must have been occupied by marsupials (Flannery 1995). In the Miocene much of New Guinea was again

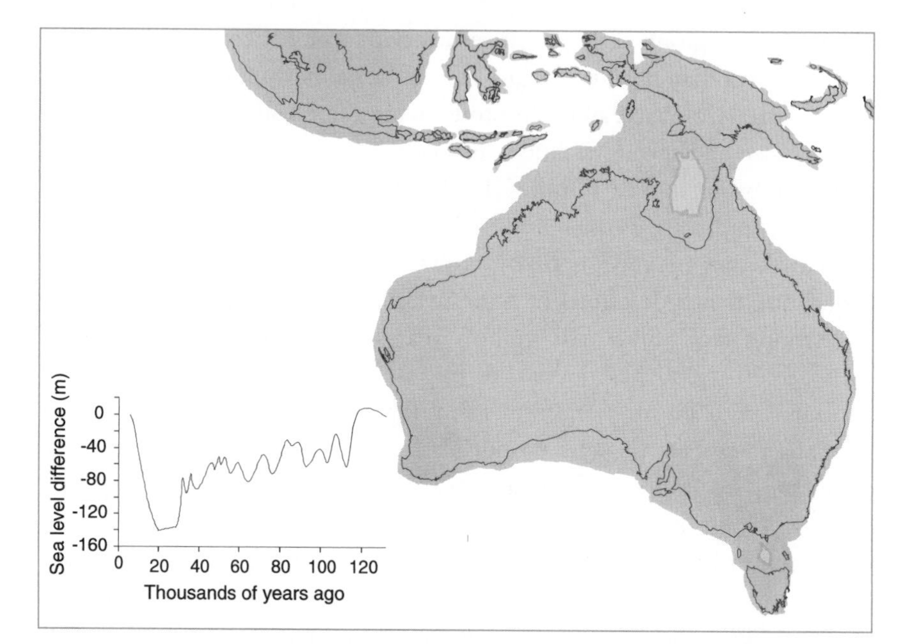

FIGURE 1.3 Map of Greater Australia at the Last Glacial Maximum (about 20 kyr ago). Pale grey areas on the Carpentarian and Bassian Plains are large lakes. The graph shows changes in sea level around northern Australia through the last glacial cycle, measured as deviation from the present-day stand (from Lambeck *et al.* 2002).

submerged, and the small areas in the central and southeastern highlands that remained as dry land were isolated from Australia. Mountain-building in New Guinea began in earnest about 10–15 million years ago as a result of the collision of the northward-moving Australian continental plate with the Pacific Plate. This gave New Guinea its modern shape and its immense topographic diversity, and also its humid climate that persisted while northern Australia dried out during the late Miocene and Pliocene. New Guinea is now the centre of diversity for some marsupial lineages such as tree kangaroos (*Dendrolagus*) and cuscuses (*Phalanger, Strigocuscus* and *Spilocuscus*) that are specialised for life in complex wet forests. The land connection with northern Australia was periodically re-established as sea levels fell during the Pleistocene glacial periods (Figure 1.3), but habitat differences between New Guinea and northern Australia meant that there was little mixing of the two mammal faunas.

MIGRANTS FROM THE NORTH

Australia has 69 species (more or less; the number is still subject to some change due to taxonomic revisions) of native rodents. All belong to the subfamily Murinae of the huge rodent family Muridae, which includes rats and mice. They have their ultimate origins to the north of Australia, and they came in relatively recently as Australia and New Guinea moved closer to southeastern Asia. Most of the native Australian murines – and all of the most distinctive and specialised of them – belong to a group called the conilurines. They are the descendants of a single lineage that entered Australia probably about 4 million years ago, or perhaps a little earlier (Breed & Ford 2006). These earliest Australian rodents took full advantage of the fact that the dry-country environments then expanding over northern and central Australia were empty of seed-eating mammals. They underwent a spectacular radiation that produced 49 species with a wide range of morphologies mainly reflecting adaptation for semi-arid and arid habitats. They include the tree-rats, stick-nest rats, rock-rats, hopping mice and pseudo-mice (Plates 3 and 4). The conilurines are endemic to Australia (with the marginal exception of one species each of *Pseudomys* and *Conilurus* that also occur in southern New Guinea, clearly as recent arrivals from Australia) and they are not closely related to any other group of rodents in New Guinea or Asia (Breed & Ford 2006).

The other Australian rodents are derived from more recent invasions, and they belong to genera that are more diverse in New Guinea. The mosaic-tailed rats (*Uromys* and *Melomys*) occur in New Guinea and in northeastern Australia and are mainly restricted to rainforests. Their distribution suggests that they entered Australia from New Guinea by way of a forest corridor over the Torres Strait land bridge, and the limited fossil evidence suggests that this happened in the early Pleistocene. The water rats *Hydromys* and *Xeromys* have rather more extensive distributions in Australia, around the coast and through inland waterways, and they could conceivably have invaded from New Guinea along the coast or along streams flowing onto the Carpentarian Plain early in the Pleistocene. Lastly, there are seven species of native true rats (*Rattus*) in Australia, two of which also occur in New Guinea (along with a larger number of endemic New Guinea *Rattus*). They tend to be associated with forests and tropical savannas, although one species, the plague rat *R. villosissimus*, occurs through central Australia.

BATS

There were bats in Australia early in the Tertiary, and a fossil from the Tingamurra fauna is one of the world's oldest known bats. It is very similar to fossils from Europe of about the same age, suggesting that bats became very widespread early in their evolution. The 79 species of bats now living in Australia belong to seven families that are widespread through the tropics. Presumably they entered Australia from the north, and there are more bats in the northern tropics of Australia than in the south. The bat families represented in Australia have quite deep fossil histories on this continent, with the puzzling exception of the Pteropodidae (the flying foxes, or fruit bats), which do not appear until the Pleistocene (Long *et al.* 2002).

The mammal fauna of Australia at the beginning of the last glacial cycle, 130 000 years ago, was thus the product of a long and complex history involving the evolution of an ancient and specialised Gondwanan fauna, the diversification of successive lineages arriving from both the south and the north, and the shaping of adaptations in those lineages by a long-term trend towards aridity and variability in the climate. The remnants of a rich and ancient rainforest fauna persisted in the wet habitats of northeastern Australia, and more fully in New Guinea. But most of the inland was populated by a dry-country fauna dominated by a highly diverse suite of herbivores and small insectivores, as well as some remarkably specialised large predators. If any of us could visit Australia before the last ice age, we would be most struck by the spectacular diversity of very large mammals. The extinction of these giants was the first great shock to be suffered by Australian ecosystems in the last 130 000 years. The next five chapters of this book investigate these extinctions, and explain why they happened.

Part I

MAMMALS AND PEOPLE IN ICE-AGE AUSTRALIA

[2.6 million to 10 000 years ago]

We live in a zoologically impoverished world, from which all the hugest, and fiercest, and strangest forms have recently disappeared; and it is, no doubt, a much better world for us now they have gone. Yet it is surely a marvellous fact, and one that has hardly been sufficiently dwelt upon, this sudden dying out of so many large mammalia, not in one place only but over half the land surface of the globe.

Alfred Russel Wallace (1876)

2

The Pleistocene megafauna

AUSTRALIA AND NEW GUINEA lost around 55 species of mammals towards the end of the Pleistocene. The species that disappeared in this wave of extinctions are collectively termed the megafauna because many of them were very large, and the event is referred to as the late Pleistocene megafauna extinction. It dramatically reduced not only the number but also the ecological range of vertebrates in Australia: the continent lost all of its large browsers, most of its large grazers, its largest predators, anteaters and scavengers, and all of its large omnivores. The loss of these species was a catastrophe for mammal diversity on this continent, but what caused it? This question is at the heart of perhaps the oldest scientific debate in Australia. The next few chapters review this debate and present the evidence needed to resolve it. But first, to better define the problem, this chapter summarises what we know of the biology of the extinct species.

AUSTRALIA'S PLEISTOCENE BESTIARY

Mammal species known to have been present in Australia and New Guinea during the second half of the Pleistocene but which did not survive into the Holocene are listed in Table 2.1, with summary information on their distribution, ecology and body size. Most of this information comes from Murray (1991) and Long *et al.* (2002), with some expansion from other sources cited below.

MEGAHERBIVORES

The extinct large mammal herbivores of the late Pleistocene can be divided into two major groups: large wombat-like creatures, including the diprotodons and at least one palorchestid (the 'marsupial tapir'), as well as true wombats and a giant koala; and a wonderfully diverse set of kangaroos, consisting of some larger relatives of living kangaroos and the completely extinct short-faced kangaroos.

The diprotodons and their relatives (family Diprotodontidae) were heavily built animals that ranged from pig-sized up to the giant *Diprotodon optatum*, which was the largest marsupial ever. *D. optatum*, a huge short-legged beast, was widespread and common through the dry open plains and woodlands of inland Australia (Figure 2.1). It was clearly a bulk feeder of coarse vegetation: it had broad molars for grinding tough plant material, a large muzzle which could manipulate big mouthfuls of foliage, and a massive skull with a large surface area for the attachment of powerful muscles that worked its lower jaw (Murray 1992). It also had very large lower incisor teeth that in some specimens show a pattern of wear on the outer surfaces suggesting they were frequently pushed into the ground, perhaps to up-root shrubs (Rod Wells, personal communication).

Fossilised gut contents collected by Stirling (1900) from a complete diprotodon skeleton at Lake Callabonna in South Australia contained mainly the leaves and stems of sclerophyll shrubs, and included pollen of chenopod shrubs (Andrew Rowett, personal communication). Tedford (1984) also commented on the association of *Callitris* (cypress pine) fruit with some skeletons at Lake Callabonna, suggesting the animals may sometimes have browsed on *Callitris*. Dr Jon Luly (personal communication) found daisies, chenopods and grasses in fossilised faeces of diprotodons from elsewhere in the Lake Eyre basin. Carbon isotope ratios in collagen from diprotodon bones collected to the south of Lake Callabonna suggest that the animals were eating C4 plants (grasses or chenopod shrubs) and hint that their food plants were growing in saline soils (Gröcke 1997). When diprotodons were alive the environment of the Lake Callabonna/Lake Eyre region was an arid mixed shrub/grass steppe with just a few scattered *Acacia*, *Callitris* and eucalypt trees (Luly 2001b; Tedford 1984). Elsewhere in South Australia diprotodon bones have isotope signatures consistent with browsing on C3 plants (non-grass herbs, and most shrubs and trees) or with mixed diets. In general they seem to have browsed in woodland habitats and to have increased their use of C4 plants in open plains (Gröcke 1997). *D. optatum* was evidently an ecologically flexible species capable of eating grass as well as a wide range of shrubs and small trees. There was at least one other species of *Diprotodon*, the smaller *D. minor*, which may have been less widespread.

Zygomaturus trilobus lived around the margins of the continent and extended to Tasmania, inhabiting more densely wooded habitats than the diprotodons. Long *et al.* (2002) speculate that it might have lived along waterways and been partly aquatic, like a hippo. It might have had short horns, like a rhinoceros. The remains of Pleistocene *Zygomaturus* are highly variable and it is possible that several species are currently lumped

Table 2.1 (*following pages*) Pleistocene mammal species of Australia and New Guinea that did not survive to the Holocene. *Notes:* Taxonomy follows Long *et al.* 2002 except for the sthenurine kangaroos, where I have accepted taxonomic changes suggested by Prideaux 2004 (including the recognition of *Troposodon* as a macropodine). Species that survived into the Holocene as dwarfed forms are not included. For species in **bold type** some fossils are thought to date from the last glacial cycle. Estimates of mass for *Procoptodon gilli*, *P. goliah*, *Protemnodon anak*, *P. brehus*, *P. hopei*, *P. roechus*, *Sthenurus andersoni*, *S. stirlingi*, *S. tindalei*, *Simosthenurus maddocki*, *Si. occidentalis* are from Helgen *et al.* (unpublished).

Family	Species	Mass (kg)	Description	Known distribution	Diet
Dasyuridae	***Antechinus puteus***	0.08	Large antechinus	Southeastern Qld	Invertebrates
Diprotodontidae	*Diprotodon minor*	900	Diprotodon	Inland eastern Australia	Browse
	D. optatum	2 700	Diprotodon	Inland Australia, widespread	Browse
	Hulitherium thomasettii	150	Diprotodon	New Guinea highland rainforest	Browse, possibly bamboo
	Kolopsis watutense	300	Diprotodon	West New Guinea, lowland rainforest	Browse
	Maokopia ronaldi	100	Diprotodon	West New Guinea highlands	'Hard ferns and grasses'
	Zygomaturus trilobus	500	Diprotodon	Widely distributed around the periphery of Australia (possibly semi-aquatic)	Browse
Hypsiprymnodontidae	***Propleopus oscillans***	40	Giant rat-kangaroo	Southeastern Australia, widespread	Omnivorous/ carnivorous
	P. wellingtonensis	?	Giant rat-kangaroo	Central NSW	Omnivorous/ carnivorous
Macropodidae					
Macropodinae	***Congruus congruus***	40	Kangaroo	Southeastern SA	Browse
	Macropus ferragus	150	Kangaroo	Inland southeastern Australia	Grass
	M. pearsoni	150	Kangaroo	Southeastern Qld	Grass
	M. piltonensis	30	Kangaroo	Southeastern Qld	Grass
	M. thor	30	Kangaroo	Southeastern Qld	Grass
	Protemnodon anak	131	Kangaroo	Eastern Australia, including Tas.	Browse
	P. brehus	110	Kangaroo	Widespread	Browse
	P. hopei	45	Kangaroo	Subalpine grasslands, western New Guinea	Grass?
	P. nombe	40	Kangaroo	Highland forest, eastern New Guinea	Browse
	P. roechus	166	Kangaroo	Southeastern Qld	Browse
	P. tumbuna	50	Kangaroo	Highland forest, New Guinea	Browse
	Troposodon minor	40	Kangaroo	Eastern Australia	Browse
	Wallabia kitcheneri	30	Kangaroo	Southwestern WA	Browse
Sthenurinae	***Metasthenurus newtonae***	55	Short-faced kangaroo	Southern and eastern Australia, including Tas.	Browse
	Procoptodon browneorum	50	Short-faced kangaroo	Southern and eastern Australia	Browse, or mixed feeder

Family	Species	Mass (kg)	Description	Known distribution	Diet
	P. gilli	54	Short-faced kangaroo	Southern Vic and SA, southeastern NSW	Browse
	P. goliah	232	Short-faced kangaroo	Southeastern Australia, widespread	Browse, or mixed feeder
	P. oreas	100	Short-faced kangaroo	Southeastern Qld, eastern NSW	Browse, or mixed feeder
	P. pusio	75	Short-faced kangaroo	Southeastern Qld, eastern NSW	Browse, or mixed feeder
	P. rapha	150	Short-faced kangaroo	Southeastern Australia, widespread	Browse, or mixed feeder
	P. williamsi	150	Short-faced kangaroo	Inland southeastern Australia	Browse, or mixed feeder
	Simosthenurus baileyi	55	Short-faced kangaroo	SA	Browse
	Si. brachyselenis	70	Short-faced kangaroo	Central NSW	Browse
	Si. euryskaphus	55	Short-faced kangaroo	Northeastern NSW	Browse
	Si. maddocki	78	Short-faced kangaroo	Southern and eastern Australia	Soft browse/ fruit
	Si. occidentalis	118	Short-faced kangaroo	Southern and eastern Australia, including Tas.	Browse
	Si. pales	150	Short-faced kangaroo	Southern Australia, widespread	Browse
	Sthenurus andersoni	72	Short-faced kangaroo	Southern Australia, widespread	Browse
	S. atlas	150	Short-faced kangaroo	Inland southeastern Australia	Browse
	S. murrayi	70	Short-faced kangaroo	Western NSW	Browse
	S. stirlingi	173	Short-faced kangaroo	Inland SA	Browse
	S. tindalei	127	Short-faced kangaroo	Inland southern Australia	Browse
Palorchestidae	***Palorchestes azael***	500	'marsupial tapir'	Eastern Australia, widespread, forests and woodlands	Browse, possibly including bark
	P. parvus	100	'marsupial tapir'	Central Qld	
Phascolarctidae	*Phascolarctos yorkensis*	16	Large koala	South Australia, central New South Wales	Browse
Potoroidae	***Borungaboodie hatcheri***	10	Large rat-kangaroo	Southwestern WA	Omnivorous

continued on page 20

FAMILY	SPECIES	MASS (kg)	DESCRIPTION	KNOWN DISTRIBUTION	DIET
cont.					
Tachyglossidae	***Megalibgwilia ramsayi***	10	Large echidna	Southeastern Australia including Tas.	Insects
	Zaglossus hacketti	30	Very large echidna	Fossils in southern WA, possibly northern Australia on the basis of depictions in rock art	Unknown (invertebrates)
Thylacoleonidae	***Thylacoleo carnifex***	110	'Marsupial lion'	Widespread	Vertebrates
Vombatidae	*Lasiorhinus angustidens*	50	Large wombat	Southeastern Qld	Grass
	Phascolomys medius	50	Large wombat	Central NSW	Grass
	Phascolonus gigas	200	Giant wombat	Widespread, including Tas.	Grass
	Ramsaya magna	100	Giant wombat	Central eastern Australia	Grass
	Vombatus hacketti	30	Wombat, similar to living common wombat	Southern WA	Grass
	Warendja wakefieldi	10	Small narrow-muzzled wombat	Western Vic, SA	Grass

together within the species name *Trilobus*. Three other diprotodontids – *Kolopsis watutense*, *Hulitherium thomasettii* and *Maokopia ronaldi* – are known from the Pleistocene of New Guinea. They were relatively small and may have been rather more specialised than the Australian diprotodontids. *M. ronaldi* lived above the tree-line and may have fed on highly abrasive alpine grasses (Flannery 1999).

The palorchestids were large tapir-like herbivores with short trunks and long mobile tongues like that of a giraffe (Plate 5). Their teeth were suited to feeding on tough plant material and they had especially long and powerful forelimbs, with big hands bearing long sharp claws. Presumably they fed by reaching up to grasp branches of small trees and pulling down foliage, which they manipulated with the trunk and long tongue. *Palorchestes azeal* is known from the forests and woodlands of eastern Australia. *P. parvus* was more widespread in the Pliocene and it (or a similar species) is known from one mid–late Pleistocene site in central Queensland (Hocknull 2005, and personal communication). Fossil material from the Darling Downs suggests that there may have been at least one other *Palorchestes* in eastern Australia during the late Pleistocene (Price & Hocknull 2005).

Late Pleistocene wombats ranged from the small gracile *Warendja wakefieldi* through the medium-sized living species and their extinct close

FIGURE 2.1 *Diprotodon optatum*, the largest marsupial ever. (drawing by Peter Murray)

relatives, to large extinct species up to about 200 kilograms in weight (Plate 6). Apart from *Warendja*, all the extinct species resembled living wombats in having massive jaws capable of delivering high compressive force to tough vegetation. Wombats have ever-growing teeth to compensate for tooth wear produced by an abrasive diet. By analogy with living wombats, the extinct species were grazers. Did they also dig burrows? All the Pleistocene species for which post-cranial material is available had limb proportions and features of the joints like those of living wombats and which are clearly adaptations for burrowing (Murray 1998). This implies that they all may have burrowed, especially when one takes into account that these adaptations would have severely restricted the capacity for long-range movement of an animal that did not centre its activities on a burrow. Living wombats minimise their energy demands, and hence their need to range widely, by spending most of their time in their burrows (Johnson 1998). The structural similarities of the extinct to the living wombats suggest that they used essentially the same strategy. If the 200 kilogram *Phascolonus gigas* dug burrows they must have been astonishing structures, and this species may well have been the largest burrowing animal ever to have lived (the largest living burrowing mammal is the aardvark, at 70 kilograms).

More than half the late Pleistocene extinctions were of kangaroos (family Macropodidae). One whole subfamily, the Sthenurinae, was wiped out and the surviving subfamily, the Macropodinae, was much reduced. Sthenurines differed from living kangaroos in having very short and deep skulls (hence 'short-faced kangaroos'), long and powerful forearms, and extreme reduction of all digits of the hind foot other than the fourth, so that effectively they had only a single large toe. Their short but heavy skulls would have given them a powerful bite, but they lacked adaptations such as high-crowned teeth and molar progression that allow the larger macropodines to cope with the tooth abrasion caused by eating grass. They were sturdily built and probably fed mainly by standing upright and stripping leaves off twigs and branches, rather than by bending to the ground to nibble smaller herbs (Figure 2.2). The outer digits of their forepaws were reduced, and the second and third digits were elongate and hook-like. These modifications of the forepaws may have helped them to pull down branches of low trees to reach foliage.

The sthenurines fall into two broad groups on the basis of their skulls, teeth and (presumably) ecology (Prideaux 2004). Species in the genera *Sthenurus* and *Metasthenurus* were rather more like macropodines than were the other sthenurines, and relative to them they had longer muzzles and more high-crowned cheek teeth. They would have had a less powerful bite than the other sthenurines, and they also had relatively broad incisors that would have increased bite size and allowed them to strip off several small leaves at once. These modifications would have helped them to feed on small-leafed shrubs like chenopods. They lived mainly through the drier inland. The geographic ranges of two species, *S. tindalei* and *S. stirlingi*, match the present-day distribution of chenopod shrublands.

The second group – the genera *Simosthenurus* and *Procoptodon* – had very short faces and small incisors. These features would have given them a very powerful bite, which together with the large surface area of their cheek teeth would have enabled them to grind very tough vegetation. Probably they fed on trees and shrubs with large tough leaves, which they manipulated with their hands much as koalas do. The species of *Simosthenurus* had more coastal distributions than *Sthenurus*, and probably lived in woodlands and open forests. One of them, *Si. maddocki*, had a set of unusual features – slender lower incisors, small cheek teeth and (evidently) a long manipulative tongue – that suggest a diet at least partly of fruit or at least of unusually soft browse.

Procoptodon species were similar to *Simosthenurus* in having very short and robust skulls and small incisors, but their cheek teeth were very like

PLATE 5 The marsupial tapir *Palorchestes azeal* fed on tough leaves and twigs and may also have eaten bark stripped from tree trunks. (painting by Anne Musser)

PLATE 6 *Phascolonus gigas*, an extinct giant wombat. Fossils of this species are often found around Pleistocene lakes such as Lake Callabonna in South Australia, and it may have been common at lake edges. (painting by Anne Musser)

PLATE 7 Marsupial lion *Thylacoleo carnifex*. (painting by Anne Musser)

PLATE 8 The Pleistocene giant musky rat-kangaroo *Propleopus oscillans*, a partly carnivorous species. (painting by Anne Musser)

FIGURE 2.2 A *Sthenurus* kangaroo pulling down limbs of a *Grevillea* to reach young shoots and flowers. (drawing by Peter Murray)

FIGURE 2.3 *Protemnodon anak* feeding selectively on *Acacia* foliage. (drawing by Peter Murray)

those of grazing kangaroos (Sanson 1982). It seems impossible that they were specialised grazers, because their small incisors would have restricted them to small bites and, as grass cannot easily be gathered with the forepaws, it is hard to imagine how they could have got enough to eat if they fed only on grass. Prideaux (2004) suggests they were mixed grazer/browsers. Their diet might have overlapped with that of *Diprotodon*, except that their upright stance would have given them access to higher foliage. *P. goliah* was the largest kangaroo, with an estimated mass of 250 kilograms and a standing height of two or three metres. The sthenurine lineage originated in dry open woodlands in the Pliocene and increased in diversity and abundance during the Pleistocene. Its evolutionary history therefore coincided with the major period of drying of the continent and the associated expansion of dry open woodlands and arid shrublands. No fossil sthenurines have been found in New Guinea.

The extinct macropodinae included species similar to living kangaroos. There were several species of large *Macropus* that probably looked like scaled-up versions of red or grey kangaroos. Like these living species they were grazers, and presumably preferred grassy woodlands and open plains. There were also several species of browsers. The *Protemnodon* species were moderately large kangaroos that differed from living species in having shorter feet and much longer and stronger arms. Their forepaws were broad, with long digits all about the same length. *P. anak* (Figure 2.3) had a particularly long neck and long slender muzzle, and in life it must have resembled a gerenuk – that very elegant long-necked antelope that feeds in African scrubs by standing on its hind legs, resting its forelegs on branches and picking foliage from the low crowns of thorny trees. Gut contents recovered with *P. anak* remains consisted of coarse twigs and large leaf fragments (Flannery 1982). One species of *Protemnodon* may have been a grazer: *P. hopei* has been recorded from high-elevation grassland in New Guinea, and Flannery (1992) noted that, unlike related species, its dentition suggests a diet of grass.

A large herbivorous bird disappeared along with these mammal herbivores. *Genyornis newtoni* (Figure 2.4), about three times the mass of the living emu, was the last of the dromornithids or 'mihirungs', a family that had been quite diverse in the Miocene and Pliocene and included the gigantic *Dromornis stirtoni*, which at about 600 kilograms was possibly the biggest bird ever (Murray & Vickers-Rich 2004). Remains of *G. newtoni* are very common, especially in southern central Australia, and it was clearly widespread throughout the drier inland. Miller *et al.* (2005) analysed carbon isotopes in a large sample of its eggshells from the shores of Lake Eyre, and compared them with emu eggshells of the same age.

FIGURE 2.4 The Pleistocene mihirung *Genyornis newtoni* standing next to an emu for scale. (drawing by Peter Murray)

This analysis suggested that *G. newtoni* fed mainly on browse, while the emu had a broader diet that included a large proportion of C4 plants as well as browse. Several other Pleistocene birds went extinct and, like the mammals, they tended to be much larger than their surviving relatives (see, for example, descriptions of the extinct giant coucal (Baird 1985) and giant megapode (van Tets 1985).

THE LION OF THE SOUTH, AND OTHER MONSTERS

Australia's largest mammal carnivore was *Thylacoleo carnifex* (Plate 7). The family Thylacoleonidae, to which this species belongs, is actually more closely related to the diprotodons and wombats than to other marsupial carnivores, but *T. carnifex* was a highly specialised predator. It had a set of long sharp first incisors that could have been used for stabbing, and third premolars that were extended to form long cutting blades. Its skull was short and robust, and a biomechanical analysis by Wroe *et al.* (2005) inferred that it had an exceptionally powerful bite – in fact, *T. carnifex* had the greatest bite force relative to body mass of any known mammal carnivore, living or extinct. In living carnivores high bite force is associated with the killing of prey well above the predator's own body mass. *T.*

carnifex's forepaws were armed with long curved claws and its thumb, which could be opposed to the other digits, had an especially long claw that would have given a tight grasp on prey. Its body mass has been debated, but around 100 kilograms seems a reasonable estimate (Wroe *et al.* 2003). If so, it was bigger than a leopard and well on the way to the size of an adult lioness.

T. carnifex was a long-limbed and powerful animal. Murray's (1991) analysis of its limb proportions indicated that it was 'principally an ambush hunter with a limited pursuit capability'. Wells *et al.* (1982) suggested that it was semi-arboreal and dragged its kills up into trees like a leopard, but Wroe *et al.* (2003) gave two reasons for doubting this. First, at 100 kilograms it was probably too heavy to have relied much on tree-climbing as part of its hunting strategy. And second, leopards use trees to avoid competition from other predators such as hyenas or lions, but *T. carnifex* was the largest mammal predator on the continent and by any standard was extraordinarily well armed. It is hard to imagine that it needed to avoid competition from anything.

The other large carnivorous mammals of Pleistocene Australia were the thylacine and a large form of the Tasmanian devil, both of which survived the late Pleistocene and will be considered in more detail later. Alongside these mammal carnivores were some giants among late Pleistocene reptiles. These were a monitor, *Megalania prisca* (Rich 1985), a crocodile that may have been partly terrestrial – the 'quinkan' *Quinkana fortirostrum* (Molnar 1985) – and the giant snake *Wonambi naracoortensis*. All were very large, although just how large has been debated (see below). Most bizarre of all were the giant horned turtles, *Meiolania* (Gaffney 1991). The best known of these is *M. platyceps* from Lord Howe Island. It was about two metres long, with horns on its head and a bony club on the end of its tail. Partial remains of two species, *M. oweni* and *M. platyceps* (or a close relative), have been found in Queensland. These appear to have been similar to the Lord Howe Island animal except that they were perhaps twice its size. They were probably herbivorous.

Three omnivorous mammals went extinct near the end of the Pleistocene. These species were ecologically very different from any living species. The rat-kangaroo *Boorungaboodie hatcheri* (family Potoroidae) was more than twice as heavy as the largest living rat-kangaroo. Its teeth suggest that it fed by crushing large hard nuts as well as perhaps being partly carnivorous (Prideaux 1999). The two species of giant musky rat-kangaroos, *Propleopus oscillans* and *P. wellingtonensis* (family *Hypsiprymnodontidae*), left very few fossils, and they may have been rare (Plate 8). The living musky rat-kangaroo *Hypsiprymnodon moschatus*, weighs only 500 grams and is restricted to rainforest, but the extinct species may have weighed up to 70 kilograms, and they lived in sparse woodlands and treeless plains extending into the semi-arid zone. Their teeth indicate that they were ecologically versatile and at least partly carnivorous. They may well have been the largest mammal omnivores of the Pleistocene, feeding on vertebrate prey as well as some invertebrates, fruit and soft foliage. *Propleopus oscillans* seems to have moved on four legs (rather than hopping, like a kangaroo), which suggests that it was probably not quick on its feet but was a capable endurance runner (Ride *et al.* 1997). The species may have filled an ecological role similar to that of dogs or foxes (although an analogy with bears might be even more appropriate).

The diversity of echidnas in Australia was reduced from three species to one. Echidnas fared rather better in New Guinea where the three species of long-beaked *Zaglossus* are still clinging to life, if somewhat precariously, and the Australian short-beaked *Tachyglossus aculeatus* remains widespread. *Megalibgwilia ramsayi* was similar in size to the living *Zaglossus* species but may have fed on insects rather than earthworms, as they do. Because we have no skulls of the metre-long *Z. hacketti* its diet cannot be inferred, but it must have fed on invertebrates of one kind or another.

There are no certain extinctions of possums in the late Pleistocene, but recent work in the Mt Etna fossil caves in central Queensland has uncovered an assemblage of small rainforest possums, some of them from a cave provisionally dated to about 280 kyr ago (Hocknull 2005 and personal communication). They include six species of ringtail possums (three each of *Pseudochirulus* and *Pseudocheirus*) and as many as five species of *Petauroides* (greater gliders). These species have not yet been named, but it is clear that they lived in a rainforest environment. They probably represent a late-surviving remnant of a specialised rainforest fauna like earlier faunas from Riversleigh in northern Queensland and living animals from New Guinea and the wet tropics of northern Queensland. Their extinction was probably a late episode of the long decline of Australia's rainforest possum fauna, and is part of a different story to the late Pleistocene megafauna extinctions.

In all, 10 families were affected by the 'megafauna' extinctions and most of them lost a large proportion, if not all, of their species. Figure 2.5 shows how these families fit into the evolutionary scheme of the marsupials. The extinctions were very strongly concentrated in the two major lineages – the kangaroos, and the wombats and their relatives including the marsupial lion – that accounted for Pleistocene Australia's high diversity of large open-country herbivores. The group of related families that included the palorchestids, diprotodontids, thylacoleonids, koalas and wombats was reduced from 20 to just four species. In contrast, the late

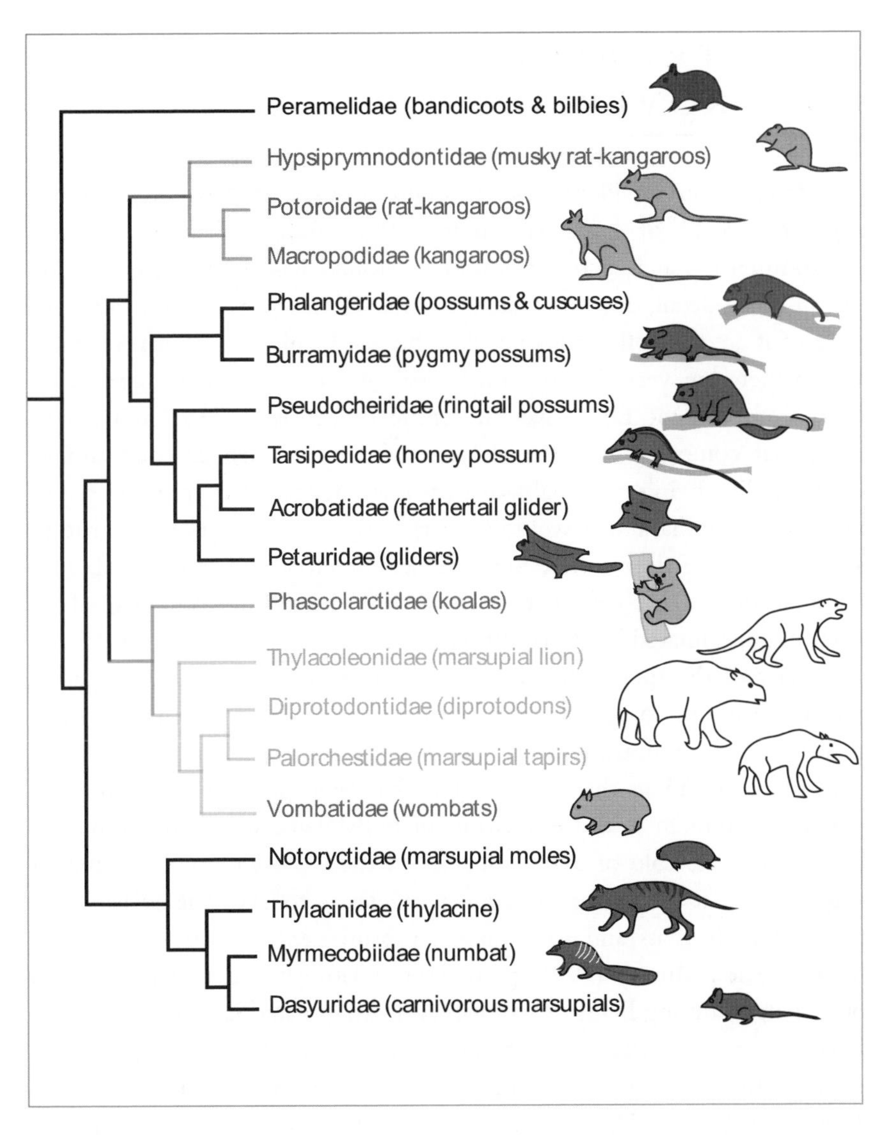

FIGURE 2.5 The phylogeny of Australian marsupials (from Cardillo *et al.* 2004 and Long *et al.* 2002), showing the evolutionary relationships of all families that were extant in the Pleistocene. Families that lost some species in the late Pleistocene are shaded in medium grey, and those that lost all species are unshaded.

Pleistocene extinctions seem to have passed over the bandicoots and bilbies, possums and cuscuses, pygmy possums, ringtail possums, honey possums, feathertail and other gliding possums, marsupial moles, thylacine, numbat and (with the exception of the loss of one *Antechinus* and the dwarfing of the devil) other carnivorous marsupials. Another way of putting this is that for smaller species that lived in trees or other complex vegetation, and that fed on insects, fruit, small vertebrates, fungi,or leaves from the canopies of large trees, the late Pleistocene extinction was a non-event. We know of no bats or rodents that went extinct in the late Pleistocene, although this could be at least partly due to the patchy fossil record for those groups.

SOME PROBLEMS AND CONTROVERSIES

The fossil record of Pleistocene Australia opens a window on an extraordinary world of large vertebrates. But while a few species and environments – such as the diprotodons around Lake Callabonna – stand out in sharp detail, our knowledge of most of this world is still hazy, and parts of it are shut off completely. The biggest blank is northern Australia, where there are very few mammal fossil sites from the Pleistocene and, apart from the Mt Etna caves in central Queensland (Hocknull 2005), none that compares with the richness of the cave deposits of southern Australia. The living mammal fauna has many species that are restricted to northern Australia and it would be surprising if this was not also true to some extent of the extinct megafauna. Therefore the true number of mammal species that went extinct in the late Pleistocene could easily have been underestimated by 20 or 30 per cent.

Many of the species listed in Table 2.1 are known only from partial remains. Without complete skeletons inferences on their feeding styles and behaviour are tenuous, and even with relatively complete material it is hard to be confident that their ecology has been properly understood. A persistent problem is the estimation of body mass, which is critical for interpreting the role of extinct species in their ecosystems. The simplest way to estimate mass for an extinct species is to take the relationship between body mass and some linear measurement of body size among living species. Murray (1991) plotted the relationship between mass and body length among living marsupials and extrapolated it to infer mass for a wide range of Pleistocene marsupials. This predicted a mass range for the extinct animals up to a maximum of a little over 1000 kilograms for *Diprotodon optatum*. This is certainly big, but far smaller than the largest Pleistocene mammals on other continents – in North America, for example, mammoths and ground sloths weighed 6000 kilograms or more.

However, many of the extinct Australian marsupials had distinctly different body proportions to their closest living relatives. Often they were more stockily built, meaning that the relationship of mass to length among living species would underestimate body mass for extinct species. Wroe *et al.* (2004b) re-estimated the mass of *D. optatum* using a method that better accounts for differences in shape. They analysed the relationship between the thickness of leg bones and body mass among living large mammals of many different kinds and sizes. The correlation between these variables is very close, presumably because it is underlain by a tight functional

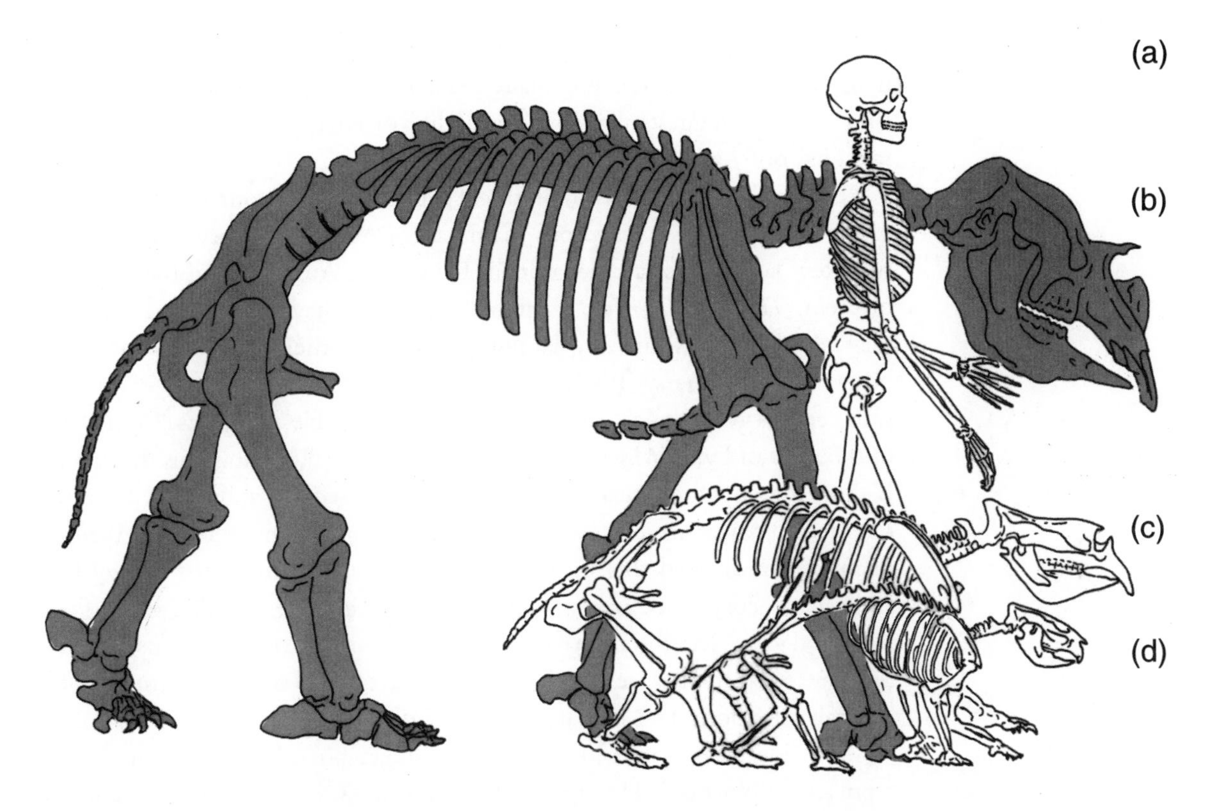

FIGURE 2.6 Relative body sizes of **(a)** human, **(b)** *Diprotodon optatum*, **(c)** the Pleistocene wombat *Phascolonus gigas* and **(d)** the living common wombat *Vombatus ursinus*. (drawing by Peter Murray)

relationship between the thickness of a mammal's legs and the mass that those legs must bear. This method put the weight of *D. optatum* at about 2700 kilograms, or somewhere between a white rhino and a female Indian elephant – these were very big animals indeed (Figure 2.6). Helgen *et al.* (unpublished) applied this method in 11 species of extinct kangaroos, and in most cases found mass values greater than had previously been estimated. It is very likely that the body masses of many other species of marsupial megafauna have been underestimated.

The question of just how heavy the biggest Pleistocene mammals were is critical to our interpretation of Australian ecosystems. Flannery (1994) argued that Australia's megafauna species were relatively small because Australian ecosystems are unproductive. In his view, the generally infertile soils and low and erratic rainfall of Australia meant that the herbivorous megafauna evolved under conditions of low energy availability, and this prevented them from attaining very large size. Burness *et al.* (2001) showed that world-wide there is a strong relationship between the area of a land mass and the size of the largest vertebrates that evolved there. They found that Australia's largest land mammal was unexpectedly small given this relationship, and interpreted this as a function of Australia's low-energy environments. But they assumed a mass of just over

1150 kilograms for *D. optatum*, whereas Wroe *et al.*'s (2004b) alternative figure of 2700 kilograms is actually greater than expected from Australia's land area and suggests that low availability of energy has not constrained maximal body size in Australian mammals.

Estimates of body mass of extinct species also affect our interpretation of the role of large carnivorous mammals in Australian ecosystems. There were few large mammal predators in Pleistocene Australia, and the biggest of them, *Thylacoleo carnifex*, is often portrayed as having been quite small relative to big cats and the largest wolves on other continents; on the other hand, Flannery (1994) argued that Australia's Pleistocene carnivorous reptiles were much larger and more formidable predators. Estimates of the mass of living *Megalania* have ranged up to 1000 kilograms or more. If these are anywhere near to being accurate, *Megalania* must have been a terrifying creature, more like a *Tyrannosaurus rex* than a goanna. Again, Flannery interpreted this in relation to the low availability of energy in Australian ecosystems. Energy constraints should be especially severe for top predators. Mammals have much higher rates of energy use than reptiles, so it makes sense that in the energy-poor environments of Australia the dominant predators were reptiles rather than mammals.

This idea was criticised by Wroe (2002), who called it 'the myth of reptilian domination'. He argued that body masses of the largest reptiles, which are mostly known only from fragmentary material, have been hugely overestimated: on his view of the evidence *Megalania* had an average mass of about 100 kilograms (although a few old individuals may have been well above this). Further, he argued that on the basis of morphology there are no strong grounds for believing that *Megalania* did not get a large proportion of its food from scavenging, like most living goannas and the komodo dragon, which it must have resembled. The fact that it left few fossils indicates it was uncommon. Even less is known about the ecology and habits of *Quinkana fortirostrum* and *Wonambi naracoortensis*, but as for *Megalania* the evidence that they were large terrestrial predators is quite weak. On the other hand, Wroe *et al.* (2003) revised the mass of *T. carnifex* up to 100 kilograms or more. This species was clearly widespread and abundant, and its anatomy leaves no doubt that it was a powerful predator. It remains true that for the last few million years at least, Australia had very few species of large mammal predators, but Wroe *et al.* (2004a) account for this as a function of the relatively small area of the Australian continent and its long geographic isolation, rather than idiosyncratic environmental features.

Perhaps the most interesting feature of the large carnivorous mammals of Pleistocene Australia is the lack not of large specialist predators, but of generalist scavengers. *T. carnifex* and the thylacine did little scavenging; the devil was a scavenger but of a very specialised kind, with jaws adapted for crushing the bones of large vertebrates. This is another indication that the dominant scavengers of the Pleistocene were the large reptiles, perhaps *Megalania* most of all. And this interpretation makes its own good sense in the light of the energy limitations placed on large carnivores in Australian environments. If prey populations were sparse, then carcasses would have appeared sporadically. A large mammal carnivore with a requirement for regular meals would therefore need to be an active and efficient hunter. Large reptiles, able to survive long intervals between meals, are the ideal scavengers in energy-poor and unpredictable conditions.

SIZE AND DESTINY

In both Australia and New Guinea the late Pleistocene extinctions were strongly size-selective, removing all species above a body mass of about 40 kilograms (Figure 2.7). The rate of extinction of New Guinean mammals (a 9 per cent loss) was much lower than for Australian mammals, but this difference is mostly explained by the fact that there were fewer large species in New Guinea. Some smaller species also went extinct, but they tended to be considerably larger than close relatives that survived. Thus the extinct late Pleistocene rat-kangaroo *Borungaboodie hatcheri* weighed only

FIGURE 2.7 Body mass and extinction of late Pleistocene marsupials from **(a)** Australia and **(b)** New Guinea (extinct species shown as white bars); **(c)** extent of size reduction from Pleistocene to living forms in relation to mean body mass of living animals, for species of Macropodids. Size change represents reduction in linear dimensions; data from Marshall and Corrucini (1978), Murray (1984), Flannery *et al.* (1996); ($F_{1,4}$ = 35.08, p = .001).

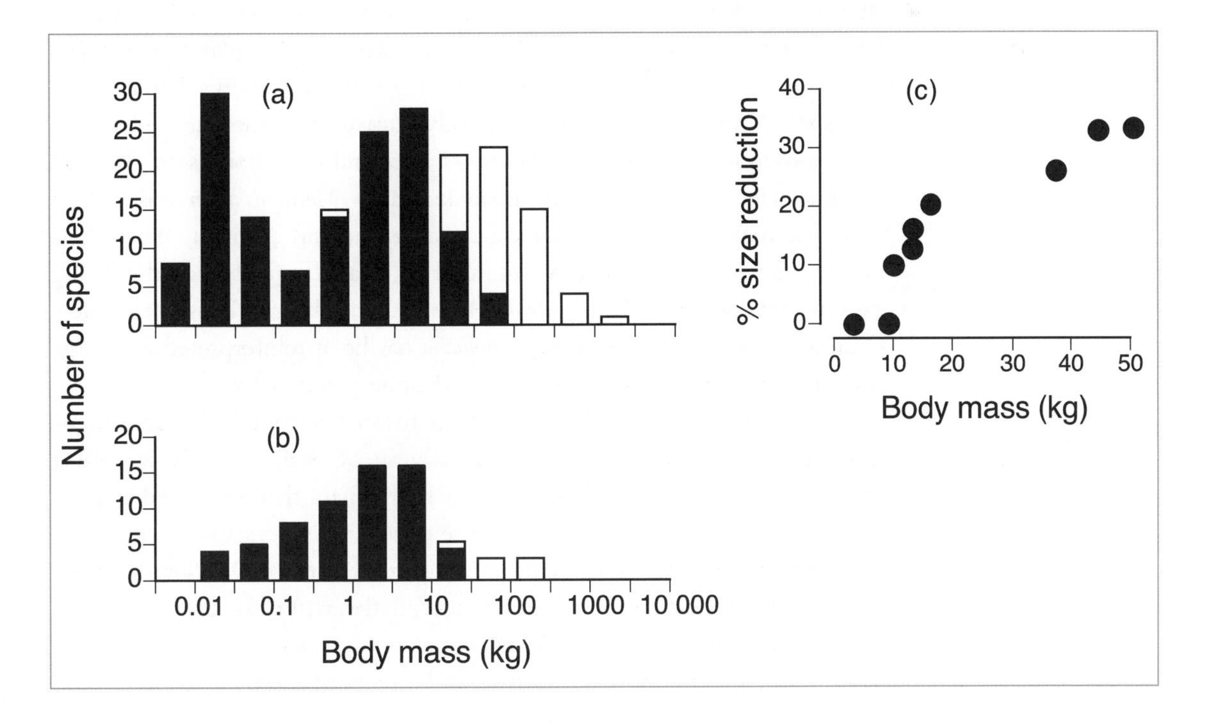

about 10 kilograms, but it was the largest member of its family; likewise, the two extinct Australian echidnas might have been small compared to the lost kangaroos and diprotodons, but beside living echidnas they were giants.

The selective extinction of larger species was a distinctive feature of the late Pleistocene. Before then most of the lineages described above had been going through a more-or-less steady increase in average body size since the Miocene. For this to be true, extinction rates of large-bodied species had to be low, at least in relation to the rate at which they arose by speciation. On the other side of the Pleistocene the few mammal extinctions recorded for the Holocene were of species that were smaller, or no bigger, than their surviving relatives. Thus the only mammal known to have gone extinct in Australia during the Holocene was the Nullarbor dwarf bettong *Bettongia pusilla,* a rat-kangaroo that was a little under half the size of living *Bettongia* species (McNamara 1997). New Guinea lost a pademelon, *Thylogale christenseni*, which was probably a bit smaller than the living *T. brunii* (Hope 1981), the smallest of the striped possums (or trioks) *Dactylopsila kambuayai*, and *Petauroides ayamaruensis*, a small relative of the greater glider *Petauroides volans* (Aplin *et al.* 1999).

Another aspect of the size-selectivity of late Pleistocene extinctions was that some species that survived became smaller; these are Flannery's (1994) 'time-dwarfs'. Late Pleistocene dwarfing has been recorded in several lineages of Australian marsupials, particularly in koalas, devils and kangaroos. (It is also possible that there was some dwarfing in wombats: Long *et al.* (2002) comment that the late Pleistocene wombat *Lasiorhinus angustidens* may possibly have been a larger ancestor of the living hairy-nosed wombats.) Quite a few species originally named from the late Pleistocene have subsequently been recognised as ancestral populations of dwarfed living species. For example, the giant kangaroo *Macropus titan* was at least twice the mass of the largest living kangaroos and was originally named as a distinct species, but it is now regarded as a large ancestor of the living eastern grey kangaroo *M. giganteus*. More recently the late Pleistocene New Guinea tree kangaroo *Dendrolagus noibano* has been reinterpreted as a large ancestral form of the living *D. dorianus* (Flannery *et al.* 1996).

Dwarfing involved reductions of up to 35 per cent in linear dimensions, approximately equivalent to a halving of body mass. It was most pronounced for species closest to the body mass that separated extinct species from survivors. For example, the eastern grey kangaroo is similar to or slightly larger than some extinct kangaroos, and the *M. titan/giganteus* lineage underwent the most pronounced dwarfing. Smaller kangaroo species, that were more comfortably below the extinction threshold to begin with, underwent less dwarfing, and species of less than ten kilograms seem not to have dwarfed at all (Figure 2.7c). It is as if there was a body-size bar in the late Pleistocene, and only species that could pass under it survived. Small species managed this with ease; some larger species ducked under the bar by evolving to smaller sizes; others were simply too big or could not dwarf quickly enough, and went extinct.

This review of the Pleistocene megafauna defines the ecological and taxonomic scope of the extinctions. It shows that the extinct species lived in habitats ranging from wet montane forests and alpine grasslands in New Guinea to dry savannas and arid steppe in inland Australia. The one feature that unites almost all of them is that they were large compared to related species that survived. However, there was no absolute threshold of body size that separated extinct from surviving species in all families, so it seems that size did not contribute directly to extinction risk. It makes more sense to think of size as conferring vulnerability to extinction indirectly, through some characteristic of species that varied with body size but did so in quantitatively different ways in different groups. The selective removal of relatively large species also applied in the birds and reptiles, with the added hint that ground-nesting birds (the giant coucal and giant megapode would have been ground-nesters, like *Genyornis*) might have been especially vulnerable.

It is these features of the megafauna extinctions – their breadth and complexity, and the contribution of relatively large size as a unifying feature – that a successful theory of cause must explain.

3

What caused the megafauna extinctions?

DESPITE A CENTURY and a half of argument, and the attention of an increasing number of specialists – in palaeontology, archaeology, quaternary environments and ecology – we seem to have come no closer to agreement on what happened to Australia's late Pleistocene megafauna. If anything, recent differences of opinion are deeper and sharper than ever.

EARLY IDEAS 1860–1968

The giant extinct marsupials of Australia first came to light in 1828 when George Rankin of Bathurst found the rich fossil deposits of the Wellington Caves on the western slopes of New South Wales (Dawson 1985). Lucky to have survived his first descent into the caves, when he tied off his rope on a projection that turned out to be a fragile *Genyornis* femur, Rankin returned to collect specimens with the geologist and Surveyor-General of New South Wales, Major Thomas Mitchell. Their material was sent to England in 1830, where it was examined by the great anatomist Sir Richard Owen and used by him in the first scientific description of an extinct species of marsupial (a *Diprotodon*).

The discovery of fossil marsupials in Australia was an important event in nineteenth-century science, because it helped to confirm what Charles Darwin called 'the law of the succession of types'; that is, the principle that each region of the earth has its own distinctive suite of fossil species that is allied with the living fauna of that region and not with other faunas. It also extended knowledge recently acquired for Europe and North America, that in many parts of the world there have been geologically recent extinctions of giant mammals. Owen included the fossil marsupials of Australia in his general views on the causes of extinction. These were that, before the advent of humanity, species went extinct through failure to 'adjust' to changes in environmental conditions produced ultimately by 'continuous slowly operating geological changes' that affected the availability of resources necessary for life. Owen thought that large-bodied species would be most sensitive to such changes because of their greater requirements for food and water. But he believed that many of the more recent extinctions had been hastened by people, and he included many of the giant mammals in this, particularly in Europe where artefacts from 'a rude primitive human race' had been found with the remains of extinct mammals (Owen 1861). He believed that people had been the main cause of Australia's megafauna extinctions, even though in his day there was no evidence that the extinctions followed human arrival, because 'No other adequate cause suggests itself to my mind save the hostile agency of man' (Owen 1877). He noted the large size of the extinct species and argued that they would have been conspicuous to human hunters, but slow-moving and unable to escape.

But the most widespread view in the late nineteenth and the early twentieth century was that a drying of the climate at the end of the Pleistocene was responsible for the extinction of the megafauna. The Pleistocene was thought to have been a cool, wet period, partly on the basis of the megafauna's existence, which implied that 'the whole country, wherever the soil was favourable, was more or less clothed with a luxuriant growth of vegetation capable of supporting these huge herbivores' (Wilkinson 1885). The transformation of this luxuriant landscape into the dry and sparsely vegetated plains of today doomed the giant herbivores:

> Stinted of their food supplies, and being unable from their great bulk to migrate rapidly or adapt themselves readily to the altered conditions of life, *Diprotodon* and the other large herbivores, perished by degrees from the combined effects of want of sustenance, the raids of predatory beasts, and possibly the attacks of man. (Wilkinson 1885)

As evidence for the dominant role of drying of the climate in the extinctions, Wilkinson referred to the Cuddie Springs fossil site, where the remains of *Diprotodon*, *Sthenurus*, *Macropus titan*, large wombats, *Genyornis* and giant Pleistocene reptiles had recently been found. This site seemed to provide evidence of the final stages in the extinction of these creatures, and pointed to the major cause:

> Nothing but want of water could have brought together such a heterogeneous assemblage of animals to the same drinking-place; and what must have been their last terrible struggle for existence, as the supply of water failed, must be beyond description.

Wilkinson's opinion on the dominant role of climate was shared by other geologists (e.g. Browne 1945). Norman Tindale (1959), the preeminent anthropologist and field archaeologist of the first half of the twentieth century, expressed doubts about the climate hypothesis on the grounds that it did not explain why so many mammal species survived in arid Australia, and why giant mammals had also gone extinct in New Guinea. He suggested that Aborigines may have eliminated some of the giant species, but then coexisted in an 'uneasy balance' with the rest of them until the arrival of the dingo 'tilted the scales towards destruction of the whole of the remainder of the Pleistocene megafauna'. However, an impact of the dingo on the megafauna had already been rejected by Gill (1955) because megafauna had gone extinct in Tasmania in its absence. Gill cast doubt on all other explanations then available, but made some additional points. He suggested that the instability of climate at the Pleistocene/Holocene boundary and during the Holocene may have been just as significant as any particular set of extreme conditions, and large animals in particular may have had difficulty adjusting to rapid swings in climate. Furthermore, as species differ in temperament, some being more resilient and tenacious than others, the extinct species may have been psychologically weak in some way that explained their sensitivity to a range of impacts. If true, this would make the causes of their extinction unknowable, because no markers of psychological failings can fossilise.

THE MODERN DEBATE

The year 1968 marks a turning point in the debate. Until then the idea that increasing aridity had caused the megafauna extinctions had come to be widely accepted, even if it was occasionally criticised, and this was the explanation given in general reviews of the Quaternary history of Australia (e.g. Burbidge 1960; Gentilli 1961). A few authors allowed a minor role for people, without developing explicit models of the nature of human impact on the megafauna. This reflected a general view that Australian Aborigines had had very little influence on their environment: they were, after all, a technologically unsophisticated people, their populations had been small and dispersed, and they had wandered through the landscape responding to natural variations in the supply of food and water rather than exerting control over resources. Two papers published in 1968 challenged this view in very strong terms (Jones 1968; Merrilees 1968): they wondered if the impact of Aborigines on the Australian environment had been dramatic, even catastrophic. The new tone can be judged from the title of Merrilees's paper, 'Man the Destroyer'.

From that point it is possible to assign many of the protagonists in the debate to one of two more-or-less equally balanced sides: those who gave the primary role to human impact and those who saw other processes (mainly climate) at work. The papers by Merrilees and Jones also distinguished indirect effects of people on megafauna due to habitat change caused by widespread burning (which they considered to have been most important), and direct effects due to hunting. This provides us with three broad categories of cause: climate change, landscape burning by people and hunting by people, all of which remain current.

THE GREAT DROUGHT

In the interval leading up to the coldest and driest phase of the last glacial cycle environmental pressures on large mammals presumably increased as Australia became more arid. Many people have suggested that this change at least contributed to the late Pleistocene extinctions (Archer 1984; Field & Dodson 1999; Horton 1984; Kohen 1995). The effects on megafauna of the deteriorating ice age climate have been conceived in two major ways. First, the major impact could have come from an increase in climate variability, as might well have accompanied the transition from one climate regime to another. Main (1978) suggested that an unstable climate would have been to the disadvantage of large-bodied species because of their generally low rates of population growth. A population of a small-bodied species knocked down by an extreme climate event, like a severe drought, might be able to recover before the next one hit; populations of large-bodied species, unable to rebound so quickly, could be driven down to very small numbers and ultimately to extinction by a series of extreme events. Main (1978) saw dwarfing as an adaptation to climate variability, because reduction in body size would be associated with earlier maturity and therefore a shorter generation time and faster recovery of populations. He suggested that the megafauna species that went extinct either lacked the evolutionary potential or were simply too large to reduce body size far enough to allow them to ride out the environmental variability of the Late Pleistocene.

Second, a general reduction in rainfall would have reduced the availability of drinking water and the productivity and nutritional quality of vegetation. Supporters of climate-driven extinction see large mammals as being most vulnerable to those changes because of their large requirements for food and water. Flood (1999) noted of the megafauna that 'The one thing they all had in common was large size and a gigantic thirst', and Bowler (1998) remarked that 'The progressive deterioration of climate in approach to the last glacial maximum ... would have imposed nearly impossible stresses on animals with large energy requirements'.

Horton (1984, 2000) has provided the most detailed account of just how these stresses might have caused extinction of large mammals. He argued that most of the extinct megafauna were species of woodland rather than truly arid habitats. Arid conditions expanded from the centre of the continent towards the coasts in the last glacial cycle, and in this process woodland habitats were compressed and fragmented around the margins of the continent. As a result, formerly large and widespread populations of megafauna were confined to small isolated refuges where they were vulnerable to local extinctions. Within these refuges declining rainfall meant fewer sites had permanent surface water, essential for large-bodied species that needed to drink regularly. As some water points dried up, the distances separating remaining water points increased until animals that depended on access to free water were unable to travel between them. Populations of megafauna thus became 'tethered' to restricted zones of habitat within range of waterholes. These zones of habitat were degraded, food supplies were exhausted by animals who for want of water could not move away to use other areas, and populations died out. The repetition of these events at many locations eventually resulted in the total extinction of species. If the intensity of seasonality or between-year variability in rainfall also increased under the harsh conditions of the LGM (Last Glacial Maximum), occasional very deep droughts would have increased the pressures on small isolated populations of large mammals.

There seems to be support for this scenario in locations where large collections of fossils of extinct Pleistocene species are found around what might have been isolated water-bodies. For example, there are hundreds of intact skeletons of *D. optatum* buried just beneath the surface of the dry bed of Lake Callabonna in the arid north of South Australia (Tedford 1984). The sediments surrounding these skeletons indicate shallow marshy conditions when the animals died, and in many cases their feet are buried most deeply: presumably they had become mired while trying to cross boggy flats during periods of low water. There are even signs of churning of the sediments around some skeletons, which tell of an animal thrashing in the mud. Archer (1984) interprets this accumulation of skeletons as representing 'one of the "last stands" of this species':

> The seeming need on their part to cross the muddy and evidently treacherous surface of a drying lake may be an indication of a non-Human critical factor in their extinction – reduction in the availability of fresh water. They may have needed to reach the shrinking body of water in the lake. This need for water, combined with the probable over-grazing of the surrounding vegetation, the difficulty of effectively replacing their numbers

when faced with unpredictable and short 'good' seasons, and possibly the presence of Man, all finally put the lid on the diprotodontid radiation.

The idea that drought caused the megafauna extinctions has been around for a long time. Horton added some ecological dimensions to the basic suggestion, but it is still vulnerable to the criticism made by Tindale in 1959 and updated by Flannery (1994). There were extinctions in all the major environments of Australia and New Guinea. While it might be possible to conjure an image of catastrophic drought causing extinction in the Australian inland, this is not so easy in the highlands of New Guinea where we know that rainfall remained high. So, what could have caused all of New Guinea's diprotodontids and large macropods to disappear along with their Australian relatives? Flannery argued that this pattern requires a mechanism that would have operated in the same way in widely differing climate zones, and for all its sophistication Horton's model cannot possibly provide such a mechanism.

Horton has two responses. First, he suggests that vegetation changes in the approach to the LGM may also have been substantial in New Guinea as a result of movement of ice caps in the highlands (Horton 1984). This implies that a rather different set of causes of extinction operated in New Guinea, associated with disruption of vegetation rather than lack of free water and local population pressures. Second, Horton (2000) points out that recent El Niño events have caused droughts in New Guinea and Indonesia as well as in Australia. By analogy with these events the extended and severe droughts of the late Pleistocene may have caused enough disruption to environments in New Guinea to trigger megafauna extinctions there as well.

Models of climate-driven extinction face another major problem. Cycles in temperature and rainfall have been a feature of world climate for the last two million years. The Australian megafauna had survived many previous episodes of aridity that must have been broadly similar to the last one. One would expect to see extinctions at the beginning of this sequence of climate cycles, not at its end. Not only that but, as Prideaux (2004) pointed out, groups like the sthenurine kangaroos had evolved under the conditions of generally increasing aridity and climate variability that characterised the whole of the Pliocene and Pleistocene. Why, after several million years of adaptation to aridity, did all the species in these groups suddenly go extinct in a drought?

There are two kinds of response to this objection. The first is to argue that the impacts of the last glacial cycle on Australian environments were significantly greater than any of the preceding cycles, as Bowler (1998), Wroe *et al.* (2004c) and others have done. Horton (1984) suggested that it was only during the last glacial cycle that the arid core of the continent expanded far enough, and the water sources in the highlands fluctuated greatly enough, to eliminate the megafauna in their habitat refuges around the margins of the continent. The second and more subtle response is to suggest that although the megafauna may have survived previous cycles of aridity, their survival was tenuous in each case. They were (quite literally) pushed to the edge of extinction each time, and it needed only a slight variation in the severity, pattern or duration of harsh climate conditions to topple them over the edge (Horton 2000).

Other criticisms can be made of climate models of extinction. For example, it is easiest to imagine how climate change could extinguish species with narrow climate tolerances or special habitat requirements, but the idea comes under strain when extended to some of the megafauna, the iconic *Diprotodon optatum* especially, that were very wide-ranging and ecologically flexible. Also, an important premise of Horton's model is that few, if any, of the extinct megafauna were adapted to deeply arid conditions. He justified this on the grounds that most of the fossils for these species are from relatively high-rainfall areas. But we have very few Pleistocene fossil sites in arid environments and megafauna are well represented in those that we do have. There is a similar problem with extinction stories based on accumulations of fossils around ancient waterholes. We find many fossils in such sites because they provide good conditions for fossilisation, not necessarily because animals concentrated around them as a result of deterioration of the broader environment.

So, there are many uncertainties in models of climate-driven extinction, and I suspect that to a large extent they are believed because there seems to be too little direct evidence for the major alternative. This reasoning is illustrated by Bowler (1998). He drew an analogy between the megafauna extinctions and the disappearance of hairy-nosed wombats *Lasiorhinus latifrons* from the Willandra Lakes region about 25 kyr ago. Before 25 kyr ago wombats were very common in the area:

> the persistence of wombat burrows in the dune sediments is clearly evident. Occurrences of articulated animal remains are frequent. Moreover, where one is exposed, further search often reveals other members of the same colony nearby. The traces of fossil burrows in the dune sediments often lead to animal remains indicating that many died in their burrows. (Bowler 1998)

But there is no indication that the animals were hunted by the human population of the area, and 'the complete absence of human agency then leaves environmental stress as the most probable cause of extinction'.

FIRE AND THE REMAKING OF AUSTRALIA

Merrilees (1968) and Jones (1968) argued that landscape burning by Aboriginal people brought about major changes to Australian vegetation, which if not the sole cause were a very significant part of the complex of factors that drove the megafauna extinct. Their arguments were based on observations of the pervasiveness of landscape burning by traditional Aborigines, as noted by virtually every European explorer. For example, Jones cited François Péron's comment in Tasmania that 'wherever we turned our eyes, we beheld the forests on fire'; travelling through central Australia, Ernest Giles (1889) wrote that 'the natives were about, burning, burning, ever burning; one would think they were of the fabled salamander race, and lived on fire instead of water'. Surely, the argument went, this level of use of fire must have changed the vegetation dramatically over practically the whole of the continent, and it may have destroyed the habitat of many megafauna species.

There is a problem with the fire hypothesis when it is stated in these broad terms. Many of the extinct megafauna seem to have been animals of open grasslands, woodlands and shrublands. When Europeans first came to Australia they found vast areas of such habitats under Aboriginal burning. If there is no reason to think that species like *Diprotodon optatum* and *Procoptodon goliah* needed radically different habitats to those that existed over most of the continent two hundred years ago, it is hard to imagine how a change in habitat caused by Aboriginal burning could have wiped them out. In many cases the most sensible view might be that burning favoured them, as it does some of the largest herbivores in Australia today. As Horton (2000) said, 'If creating grasslands in Australia was good for cows and sheep [it] would have been just as good for the Diprotodons'.

A more subtle version of the hypothesis is as follows. Before the arrival of people many of the habitats of inland Australia had a complex layer of shrubs and small trees, which was important to the large number of megafauna species that were browsers. An increase in burning removed or simplified this vegetation layer over large areas. Miller *et al.* (1999) suggested that the giant browsing bird *Genyornis newtoni* may have gone extinct as a result of such a change. Archer, Hand & Godthelp (in McGowran *et al.* 2000) also pointed out that the loss of so many browsers implicated vegetation change as a cause of the extinctions, and fire may plausibly have caused vegetation change. Bowman & Prior (2004) wondered if Aboriginal burning might have caused a rearrangement of habitat mosaics, creating a fine-scale patchiness of shrub and tree communities that disadvantaged large browsers but favoured grazers and small mammals. These ideas have not been extended to all megafauna species or to all habitats, and it is hard to imagine how they could be adapted to account for the extinction of rainforest diprotodontids and macropods in New Guinea. Nonetheless, landscape burning continues to be invoked in general terms as a factor that might have contributed to at least some of the extinctions.

OVERKILL

The idea that people hunted Australia's megafauna to extinction was the starting point for discussion of the problem in the nineteenth century, and practically all writers on the topic have been willing to accept that people did hunt the extinct species to some extent. The challenge has been to identify a convincing scenario in which hunting by a (presumably) small human population could have caused the extinction of so many species. Two kinds of mechanism have been proposed for this, often referred to as 'blitzkrieg' and 'attrition'. Blitzkrieg (literally 'lightning war') envisages very rapid extinction as a result of massive overhunting very soon after human arrival. Attrition is the slower decline to extinction of populations under a regime of low hunting intensity. Defined like this, the two models are distinguished primarily on the pace of extinction. But it is difficult to specify just how fast extinction should have been to qualify as a blitzkrieg, and a more useful distinction is between (1) hunting of megafauna as a specialised activity which provided a large proportion of food for early human populations and (2) hunting as a relatively minor component of a broad hunter-gatherer economy. In the first case, the rate of growth of the human population would have depended strongly on the availability of megafauna prey; in the second, human growth rates would have been insensitive to availability of megafauna. Extinction under the first of these assumptions equates to the classic formulation of the blitzkrieg hypothesis in North America by Paul Martin (1973).

Martin wanted to explain the disappearance of large mammals – mammoths, giant deer, camels, horses, ground sloths, gomphotheres, glyptodonts, and others – soon after the colonisation of the Americas by people about 11 kyr ago. He argued that the first Americans found a land in which very large mammals were diverse, widespread, abundant and – crucially – naive to human hunters and therefore easy to kill. The naivety of the megafauna combined with their great size meant that the hunting of them was a very profitable activity. People specialised in giant-mammal hunting and the human population increased very quickly as a result. This in turn meant that the predation pressure on the large mammals became very great, and within a short time their populations crashed to extinction. People then had to turn to alternative food sources or move on to find more megafauna to kill. This process took the form of a wave that, driven onward by the migration of people from places where megafauna had been overexploited into areas where they were still abundant, travelled from one end of the Americas to the other within a few thousand years. The rapidity of extinction at the wave-front is an essential element of the model, because a long period of interaction of people with megafauna would have allowed the initially naive creatures to learn or evolve appropriate defensive or escape behaviours, and so reduce their susceptibility to hunting.

Flannery (1990, 1994) proposed that very similar events happened in Australia and New Guinea. If anything, he implies that in this part of the world there was an extreme version of the process, because before people arrived the main predators of the Australian megafauna were large reptiles. Their lack of any evolved response against mammalian predators could have meant that Australia's megafauna were especially slow to react to the threat from people. And like Richard Owen before him, Flannery (1994) pointed out that many of the extinct species were probably slow-moving and would have been poorly equipped to escape from human hunters even if motivated by the appropriate fear responses.

An alternative view is that people hunted megafauna but did not specialise in large-animal hunting, which may actually have been a rather small component of a broad hunter-gatherer economy. Murray (1991), Johnson (2002) and Prideaux (2004) have suggested that light hunting of this kind could have caused the extinction of large mammals, which were vulnerable to it for two reasons. First, because very large mammals typically have low reproductive rates and long development times, the rate at which animals killed by hunters can be replaced is very low. Therefore a small addition to natural mortality by hunting could hold death rates above birth rates and cause population decline in very large species, while for smaller species a similar rate of mortality from hunting would be compensated by increased reproductive success. Second, large mammals typically have naturally low population densities, so that a small absolute number of animals killed by hunters per year translates into a demographically significant mortality rate.

Could this kind of low-level hunting have driven species all the way to extinction? In principle, yes, if all individuals in the prey populations were potentially exposed to hunting at some time in their lives. This would be most likely if hunters used all of the habitats in which the prey species lived and there were no refuge areas in which the animals were invulnerable. If human population size did not depend strongly on megafauna hunting (because people were also using a variety of other resources) there would still be plenty of hunters around after megafauna populations had declined to critically low levels, so individuals in these low-density populations would still be at risk of encountering people and being killed. This model does not require that the extinctions were very rapid or assume that the mammals that went extinct were any more naive to hunters or slow-footed than those that survived; only that they were demographically more sensitive to small increases in mortality imposed by hunting.

The two models outlined above are not strict alternatives, but rather they represent the ends of a spectrum of possibilities that differ in the weight given to features such as specialised hunting of megafauna and prey naivety. The emphasis placed by Murray (1991) and Johnson (2002) on the demographic vulnerability of large mammals is clearly also helpful in accounting for the abrupt collapse of megafauna populations envisaged by Flannery.

THE FLANNERY EATERS

Of all the proposed causes of megafauna extinction, overkill has been most heavily criticised. This is in spite of the fact that it would seem able to provide a better explanation of the pattern of extinction than the major alternatives – climate change and fire – because it can account for the fact that an essentially similar suite of species, all of them relatively large-bodied and vulnerable to hunters, disappeared at about the same time from regions with very different climates and vegetation. Most of the criticism has been directed at the full-blooded blitzkrieg version of overkill promoted by Tim Flannery, which has been attacked at the following three points.

DUBIOUS ASSUMPTION: MEGAFAUNAL NAIVETY In Flannery's hypothesis the extinct megafauna were easy to kill because they did not react to humans as predators. This explains both why people were able to hunt them at the very high intensities needed to produce rapid population decline, and why none escaped the 'lightning war'. In arguing for the naivety of the extinct species, Flannery drew heavily on recent examples from island faunas. In places like the Galapagos Islands, for example, native species were pitifully vulnerable and would scarcely move aside from approaching humans (who in many cases had come ashore to kill them). But these are examples of species and places that had no vertebrate predators at all. All continents have (or had) large vertebrate predators, for which other vertebrates were prey. Critics of 'blitzkrieg', like Grayson & Meltzer (2003) in North America and Wroe *et al.* (2004c) in Australia argue that it is therefore not valid to use the observed naivety of vertebrate prey on islands to explain late Pleistocene megafauna extinctions on continents. In any case, Wroe (2002) stresses the point that Australia's megafauna had significant mammal predators before the arrival of people. So, the assumption of naivety of the extinct species may be unsound. The real problem is that we will never know one way or the other, because there is no possibility of extracting this kind of behavioural information from the fossil record. The point will be forever moot and, to the extent that Flannery's version of events relies on the assumption of prey naivety, it is unverifiable.

There is another reason why some critics dislike the assumption that Australia's megafauna were naive: they take offence to it because it seems to imply that the extinct species were exceptionally slow-witted, and they see in this a reflection of the nineteenth-century view that marsupials are inferior to other mammals. Horton (2000) ridiculed this aspect of Flannery's hypothesis as 'the dopey diprotodon suggestion' and insisted that 'there is no evidence that Australian animals are inferior to other animals in intelligence or adaptation to the environment'. This concern is misplaced. Martin, Flannery and others have argued that naivety to human hunters was a feature of extinct large vertebrates in many parts of the world, not a peculiarity of Australian marsupials.

LACK OF ARCHAEOLOGICAL EVIDENCE The evidence for human hunting of extinct megafauna is slight and, at best, ambiguous. There are no unmistakable 'kill sites' with evidence of the systematic slaughter and use of many individuals of the extinct species. Worse still, there is no evidence that Aboriginal people had weapons typically associated with big-mammal hunting during the period that megafauna disappeared: spear-throwers appeared perhaps 15 kyr ago (the exact date is uncertain, see Chapter 8) and stone spear-points about 4 kyr ago (Mulvaney & Kamminga 1999). Beforehand it seems that people used only simple wooden implements for hunting. These might have been effective in killing the occasional giant marsupial but seem inadequate to the task of wiping out over 50 species from a continent. The problem of lack of support from archaeological evidence – or, on some interpretations, the conflict with an archaeological record that shows that hunting of megafauna must have been on a minor scale – is therefore central to most critiques of extinction by overkill (see in particular Wroe *et al.* 2004c).

There are ways to reconcile the lack of kill sites with rapid extinction by over-hunting. As originally argued by Martin (1984), if the extinctions really were very rapid then little archaeological evidence would have been left behind. In Australia the great antiquity of the event could also mean that very little of this evidence has survived. And much of the killing might have been done by mobile hunting parties operating far from permanent occupation sites.

These suggestions are plausible enough, but to some archaeologists they border on glibness, for the following reasons:

1. If the blitzkrieg was very rapid then it must have required very efficient hunting technology. To invoke the speed of the process as an explanation for its archaeological invisibility therefore seems to require archaeological evidence of specialised weaponry. In the places where a strong case has been made for extinction of late Quaternary large vertebrates by blitzkrieg we have either many kill sites (of moa in New Zealand) or weapons clearly designed for hunting large vertebrates (the 'Clovis' points in North America), and these have been used as evidence of heavy hunting of extinct species. But in Australia we have neither.
2. If the extinctions were completed quickly, the rate of killing must have been correspondingly high. Therefore if one imagines the interval as being very short, this implies a very high potential for the formation of archaeological evidence within that time – and the lack of such evidence remains a problem.
3. It is easy to imagine that carcasses of the very largest species were not carried back to permanent camps, but less easy to accept that partly dismembered remains of large species, or of whole carcasses of smaller megafauna species, were not returned to occupation sites where they could be discovered by archaeologists.

When Flannery's 1990 paper advocating blitzkrieg in Australia was published in *Archaeology in Oceania* the journal also printed responses from eight scholars including six archaeologists (Richard Wright, James O'Connell, David Horton, Donald Grayson, Sandra Bowdler and Athol Anderson), all of whom were sceptical of the idea that human hunting had made a significant contribution to the extinctions. The three Australians in this group (Wright, Horton and Bowdler) were the most deeply opposed. Bowdler was particularly unhappy with 'that terrible old argument' that the hunting to extinction of megafauna could be archaeologically invisible. Their scepticism has also been expressed strongly by the palaeontologist Michael Archer (quoted by Nolche 2001):

> Is there a single [megafauna] skeleton in Australia that shows that a human being killed it? The answer as far as I know is 'No!' Are there any signs of the kinds of weapons that the North Americans used to slaughter the large game in North America? The answer again is 'No!' They're fairly modest tools. We can't even find the gun, *let alone* the 'smoking gun'.

One could argue that the marsupial megafauna were so naive to hunters that large numbers of them could be killed with the 'fairly modest tools' available to people at that time, but this response is unsatisfactory because it draws the whole debate into a circle around the untestable assumption of megafauna naivety.

LESSONS FROM PREDATOR–PREY THEORY When a prey species is very abundant it is comparatively easy for hunters to inflict a high mortality rate and to drive numbers down. Once that species becomes rare, however, it becomes harder to find; for the hunter this means that the effort devoted specifically to hunting it becomes increasingly unrewarding. Hunters of large mammals should therefore have turned their attention to other more abundant types of prey as the large species declined. The relief from hunting should have allowed megafauna populations to stabilise at low numbers or begin to increase again. This mechanism should allow the long-term coexistence of prey and hunters, and it explains why attempts to exterminate introduced large mammals in recent history have almost invariably failed. Choquenot & Bowman (1998), Grayson (2001), Horton (2000) and others have criticised the Flannery/Martin model for its assumption that the rate of killing of megafauna stayed high as they became rare. Choquenot and Bowman suggest that this represents a hidden assumption that megafauna were hunted not for subsistence but for some other reason (such as for social prestige or to obtain ritual objects), which meant that their value increased as they became rare. Another possibility is that the megafauna were so hopelessly naive that it was possible for hunters to inflict high mortality rates even after they had become very rare. This is plausible in itself, but the argument increases the weight that must be borne by the problematic assumption of megafaunal naivety.

HYPERDISEASE?

One more distinct hypothesis should be mentioned. MacPhee & Marx (1997) developed a very detailed proposal that prehistoric extinctions in many parts of the world were caused not by direct effects of people on fauna or environments, but by diseases that they (or their commensals) introduced to susceptible native species. This idea has the advantage that it can explain why extinctions followed human arrival in places like Madagascar and Australia where there is little evidence of hunting of megafauna. But it has been strongly criticised (Alroy 1999).

The main problems with it are the extreme characteristics that must be assumed for the disease organisms. These must have been benign to humans and capable of surviving without access to susceptible hosts, but unspeakably virulent in a very wide range of other species. We must also accept that diseases with these characteristics could have been spread to all of the places that suffered prehistoric megafauna extinctions. No recent analogues of such an event have been found (Lyons *et al.* 2004). In Australia there is a special problem in that Aboriginal people, and any diseases they might have brought to Australia, came through southeast Asia, which still has its Pleistocene megafauna of elephants, rhinos, tapirs and large primates. The disease hypothesis therefore requires that the pathogen that devastated large marsupials (as well as reptiles and birds) in Australia must somehow have spared these Asian megafauna.

OVERVIEW

The long controversy over Australia's megafauna extinctions has been quite evenly balanced. Two ideas – drought caused the megafauna extinctions; the megafauna went extinct because they were very easy for people to kill – have been central to the debate from the beginning. These ideas have been updated, augmented, refined and tested against a growing body of evidence by several generations of scientists. Nonetheless the Australian scientific community remains divided on the issue. Many Australian archaeologists are unconvinced that overhunting was a cause of the extinctions, and they are especially wary of overhunting by blitzkrieg. The three most respected texts in Australian prehistory and archaeology published during the 1990s – Flood (1999), Lourandos (1997) and Mulvaney & Kamminga (1999) – all concluded that human hunting could not have been the primary cause of the megafauna extinctions. Mulvaney and Kamminga wrote that 'the "blitzkrieg hypothesis" has lost attraction to most scholars, excepting Flannery'. They based their own scepticism on empirical grounds, but they also cited ideological objections: 'More recently, archaeologist Stephanie Garling has challenged the hypothesis, which she describes as an androcentric hero tale of "man the mighty predator"'. To most of these writers, climate change is the most plausible alternative to hunting.

Palaeontologists have more mixed views, with some like Tim Flannery, Peter Murray and Gavin Prideaux favouring overhunting as the sole or major cause of extinction and others such as Stephen Wroe and Michael Archer putting their hands up for environmental change. The most authoritative, comprehensive, up-to-date (and beautiful) text on Australian prehistoric mammals kept all options open by remarking that the late Pleistocene megafauna vanished 'probably because of human hunting, burning and/or land alteration, but possibly also because of climate change' (Long *et al.* 2002); otherwise the topic was left alone, presumably because the authors thought it intractable (or because they could not agree among themselves).

Recent dating studies by Miller *et al.* (1999) and Roberts *et al.* (2001b), which concluded that the extinctions were concentrated in a short period between 50 and 45 kyr ago, have been widely interpreted as pointing to people rather than climate as the cause, and this interpretation has been accepted in at least one new archaeology text (Morwood 2002). However, Roberts *et al.* were careful to say that their results were consistent with either hunting or fire-induced habitat change as the factor that drove the extinctions, and Miller *et al.* (1999) explicitly favoured fire.

In short, it is not yet possible to glimpse a consensus emerging in the literature on this most contentious subject. I suspect that the most widespread view is actually that none of the mechanisms described above is sufficient of itself to account for all of the extinctions (or at least that the evidence does not yet justify such a view), and that the disappearance of the megafauna must have been due to a combination of causes.

COMPLEX HYPOTHESES

The hypotheses described above can plausibly be combined in many ways. For example, Flood (1999) concluded that the extinctions were completed in two stages: the initial impact of hunting, augmented by use of fire, reduced megafauna abundance and caused some species to go extinct, and climate deterioration later completed the extinctions. Bowman (2003) suggested that new patterns of landscape burning introduced by Aboriginal people might have changed megafauna habitats, but noted that effects of this on megafauna populations might have been amplified if people used fire as a tool in the hunting of megafauna.

It is also conceivable that different factors, or combinations of factors, operated on different species or in different places. Perhaps fire contributed along with human hunting to extinctions in the dry inland of

Australia and was especially severe in its effects on browsers, but in the wet highlands of New Guinea rapid shifts in vegetation were more important. Many such complex scenarios could be constructed, increasing several-fold the number of hypotheses available. Another layer of complexity can be added by imagining that some external factor, like human impact or climate change, caused some species to go extinct, and the loss of those species then upset ecosystem processes in ways that led to further extinctions. For example, the disappearance of important prey species may have caused the extinction of specialist predators, or the extinction of some large herbivores may have led to changes in vegetation that took away the habitats of many other species (Owen-Smith 1999).

Multi-causal hypotheses have been attractive to many scholars. This is mainly because each of the three mechanisms described above can be made to sound plausible, and if one therefore accepts that they might all have been at work it is sensible to believe that they all contributed to the extinctions. It can also seem that to invoke a single process to explain all the extinctions is simplistic, and that we should be trying to construct more complicated (and realistic?) models that incorporate multiple causes and interactions between processes. White & O'Connell (1982) presumably had this in mind when they described Martin's blitzkrieg model as 'crudely unicausal'.

Although it might seem sensible to believe that the causes of the late Pleistocene extinctions were complex, I think we should begin the search for a cause of the extinctions with simple single-cause hypotheses. The reason for this is Occam's Razor, the principle that a simple model that explains a given phenomenon is preferable to a more complex explanation because it depends on fewer assumptions and can more readily be generalised to new cases. Occam's Razor is helpful to progress in science, because if a hypothesis is wrong, it is easier to prove it so if it is simple than if it is complex. Simple hypotheses can be used to generate clear predictions that are distinct from the predictions of alternative hypotheses. If a hypothesis makes predictions that do not match reality, it can be rejected. But the predictions of complex hypotheses are intrinsically unstable: a complex model that seems to be failing can too easily be rescued by adding a few more variables, or adjusting the weighting given to some of its elements. In the end it would be possible to devise a complex scenario to explain any particular extinction, but with no generality and no testability (and therefore no scientific value) for any of the models.

So, if we begin the enquiry by testing simple models we have some hope of finding a clear and intellectually rigorous path towards complex models if they really are necessary, but if we prefer complex models at the outset it is much less likely that we will ever be able to work our way back to a simple model. This is the approach that I take; if it fails, we will know that we must retreat to the messy realism of a complex model or, worse still, a range of models tailored to particular cases. Before attempting this, however, we have to think about how we can test hypotheses on the causes of extinctions that might have happened over 40 kyr ago. In fact, we need to develop some confidence that such testing is even possible.

EXPLAINING EXTINCTION

In their influential textbook on conservation Caughley & Gunn (1996) reviewed a series of case histories of species that have declined to extinction or near-extinction in recent history, beginning with the dodo. In a surprisingly high proportion of cases (including the dodo) it is not clear which factor among several possibilities was the cause of population decline. This makes the attempt to find the cause of the late Pleistocene extinctions look hopeless. If it can be so difficult to determine the causes of declines and extinctions from the recent past, it must be next to impossible to do so for the late Pleistocene. But one of the goals of this book is to find out what did cause those extinctions. In this section, therefore, I develop a framework that will allow testing alternative hypotheses with the limited evidence available to us.

There are three distinct ways in which it is possible to test if some factor might have been the cause of long-past extinctions. They consist of asking the following questions:

Was the factor correlated with the extinctions in space and time? We want to know if the extinctions actually happened at the times and in the places where the hypothesised factor was operating, and not at other times and places.

Does the factor explain why certain species went extinct and others survived? Here, we need to know that the species that went extinct had features that should have made them vulnerable to the hypothesised factor, while survivors that were also exposed to the same factor differed in ways that should have made them less sensitive to it.

Could it have been powerful enough to account for the magnitude of the extinction event? To answer this question we need to combine realistic estimates of the levels of the proposed factor with theoretical formulations of the way in which megafauna populations should have been affected, and show that extinction on the scale observed was a predictable result.

These three approaches are independent, in the sense that each can be applied in the absence of any knowledge of the others. Further, they use very different kinds of empirical data and conceptual tools: (first approach) studies of extinction chronologies, from the application of geochronological methods in palaeontology and archaeology; (second approach) comparative organismal biology, using data and models from ecology and physiology; and (third approach) theoretical modelling, grounded in evidence from archaeology and palaeoecology of the conditions that applied at the time the extinctions happened. Applying this framework to any particular hypotheses therefore guarantees the systematic use of all relevant data and conceptual resources.

In testing hypotheses on the megafauna extinctions I intend to work to this standard: if a hypothesis passes all three tests it may well be true, but if it is clearly inconsistent with any one test then it should be rejected. This is a stringent standard because it draws on all relevant empirical data, applies the maximum intellectual firepower to each hypothesis, and provides three different ways to reject a hypothesis but only one (a 'yes' to all three questions) that qualifies a hypothesis as a plausible account of what might really have happened. Even if we fail to come up with a convincing answer, applying this method should allow us to reject hypotheses that do not work. For a field knee-deep in unresolved hypotheses, this in itself would be excellent progress.

4

Two dating problems: human arrival and megafauna extinction

THE AUSTRALIAN ENVIRONMENT changed in many ways during the last glacial cycle. People arrived, the megafauna disappeared, there was an increase in fire, the climate became colder and drier, and the character and distribution of the vegetation altered dramatically. We want to know how these events were related to one another, and for this an essential first step is to find out when they happened. This chapter examines the timing of two critical events: the arrival and spread of people in Greater Australia, and the extinction of the megafauna. Fixing dates on these events has been a major preoccupation for archaeology and Quaternary science in Australia, and it has given rise to a series of controversies that are still very active. In both cases, however, I think a consensus could be in sight. My goal in this chapter is to explain enough about the evidence concerning each event that it will be possible to identify where that consensus view might rest.

FIXING DATES ON THE PAST

Dating human arrival and megafauna extinction depends on measuring either the burial age of artefacts and fossils, or determining directly the age of the materials themselves. The same suite of techniques has been applied to both the human and megafauna histories, and the most important ones are briefly described below.

Radiocarbon dating The carbon isotope ^{14}C is generated in the atmosphere by the effect of cosmic radiation on nitrogen. It enters living systems as $^{14}CO_2$ and then gradually decays to ^{14}N. A measurement of the ratio of residual ^{14}C to the stable isotopes ^{12}C or ^{13}C in a sample gives an indication of the time elapsed since the organic material in the sample was formed. The materials most often dated by radiocarbon in Australian studies are charcoal and shell. Bone can be dated directly using the radiocarbon technique, but only relatively young material retains sufficient carbon for this. The accuracy of radiocarbon dating declines with the age of the sample, as the quantity of residual ^{14}C becomes very small. In the early days of the technique it could be used only on material up to about 35–40 kyr old (the 'radiocarbon barrier'). Two sets of technical developments introduced from the mid-1980s to the mid-1990s pushed the radiocarbon barrier out to about 50 kyr. Counting of carbon isotopes by Accelerator Mass Spectrometry (AMS) took over from determination of radioactivity as the technique for measuring ^{14}C concentration and made it possible to measure very low ^{14}C values; and there were improvements in the chemical pretreatment of samples to remove contaminants, the most effective of which is a process known as ABOX-SC (acid/base/dichromate oxidation and stepped combustion).

The concentration of ^{14}C in the atmosphere has varied through time, and a calibration is needed to convert the 'radiocarbon age' of a sample to a calendar age. I present all radiocarbon ages as true calendar ages (rounded to the nearest 500 years for the Pleistocene). For age conversion I used the tools developed by the University of Cologne Radiocarbon Laboratory: the conversion program CalPal (Weninger, Joris and Danzeglocke, at www.calpal.de/) and an on-line age converter (www.calpal-online.de/).

Luminescence dating measures the age-dependent accumulation of electrons in minerals, especially quartz. The electrons are those displaced by ionising radiation (from radioactive elements in the surrounding material, and cosmic radiation) and trapped in defects in the crystal lattice of the mineral. Trapped electrons can be quantified by measuring the luminescence they produce when they are liberated either by heating the material or by exposing it to light, referred to as thermoluminescence (TL) and optically stimulated luminescence (OSL) respectively. The luminescence measurement indicates the time elapsed since the electron traps in the material were last emptied by exposure to heat (TL) or sunlight (OSL). Measures of luminescence can be converted to age if the size of the ionising dose delivered to the material per year is known; this requires an on-site measurement of the radiation dose rate. These methods are wonderfully sensitive: OSL can be used to determine the time since individual sand grains in a layer of sediment were buried. Luminescence techniques have an upper age limit (for quartz) of about 200 kyr. An alternative method of measuring trapped electrons is Electron Spin Resonance (ESR). This can be used to date bones and teeth, but it may be confounded by the problem that the post-depositional uptake of uranium by those materials (see below) results in changes to the radiation dose rate through time.

U/Th dating makes use of the decay of radioactive ^{238}U to the long-lived daughter isotopes ^{234}U and ^{230}Th. In materials that take up a dose of

^{238}U at formation, and then neither gain ^{238}U with age nor lose it except by radioactive decay, the accumulation of these isotopes provides a clock that gives reliable ages up to around 350 kyr. The method could be used for the dating of bones and teeth, except that these materials do take up uranium after they have been laid down in sediments. The application of U/Th in the dating of bone will depend on the development of methods to quantify and correct for U uptake. If these could be routinely applied they would decisively resolve some of the uncertainties of late Pleistocene chronology discussed in this chapter.

In the technical literature, measurements of age are accompanied by estimates of uncertainty. Uncertainty typically increases with the age of the sample, and may be as much as 10 per cent of the estimate. In that case the true age of a sample estimated to be 40 kyr old could plausibly be 44 kyr, or 36 kyr. For the sake of clarity I have not reported estimates of uncertainty, but readers should be aware that they are the grains of salt that should be added to conclusions on the timing of crucial events.

THE BEGINNING OF AUSTRALIAN PREHISTORY

ARRIVAL

The ancestors of the first Australians left Africa for the Middle East between 120 and 80 kyr ago. From there, they seem initially to have dispersed eastwards within the tropical belt, probably moving through the open grasslands and savannas of Asia and eventually reaching southeast Asia (Finlayson 2005). They may then have travelled from Java along the Lesser Sunda Island chain to Timor, and finally crossed the Timor Sea to northwestern Australia (Figure 4.1; Morwood 2002; Mulvaney & Kamminga 1999). The low sea levels of the last glacial meant that many of the islands along this route were connected by dry land, but the passage involved a voyage of perhaps 100 kilometres across the Timor Sea. The colonisation of Australia was a most significant event in world prehistory. It was humanity's first major sea crossing, and it was the first time that modern humans set foot on a continent that had not been occupied by any other hominids (other species of *Homo* had been living in Europe and Asia for more than a million years). When did it happen?

During the 1960s and 1970s the earliest dates for human occupation of Greater Australia were steadily pushed back to around 35–40 kyr ago, where they stuck at the radiocarbon barrier. Refinement of radiocarbon

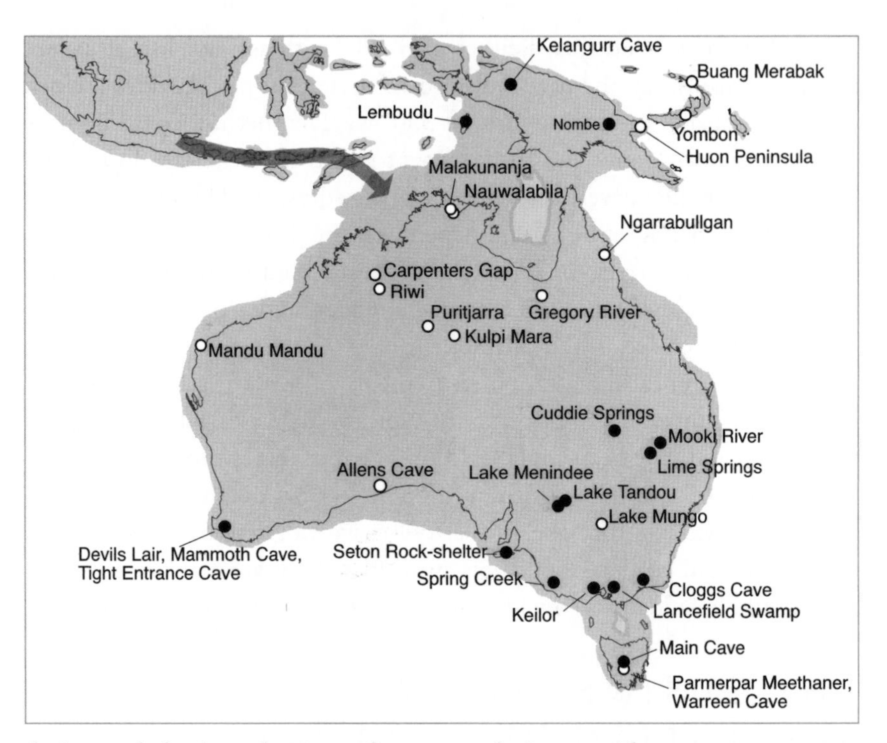

FIGURE 4.1 Map of Greater Australia, showing the locations of archaeological sites (○), and megafauna fossil sites (●) discussed in the text. The arrow marks a likely route by which people completed the migration to Australia.

dating and the introduction of newer techniques with greater age ranges stimulated a fresh round of dating activity from about 1990, and since then the needle on the time gauge for human occupation of Australia has swung around violently. It briefly touched 120 kyr with a claim for the antiquity of Jinmium Rock-shelter in the Keep River district of the Northern Territory, bounced back when that measurement was shown to be technically flawed, then swung out again when Thorne *et al.* (1999) measured an age of more than 60 kyr on a human skeleton at Lake Mungo. That study was also criticised on technical grounds (Bowler & Magee 2000; Gillespie & Roberts 2000; Grün *et al.* 2000) and, according to a precisely dated stratigraphy of the site published soon afterwards, the skeleton is only about 40 kyr old (Bowler *et al.* 2003).

The current state of the field is summarised in Table 4.1, which lists all archaeological sites from Greater Australia for which dates of 40 kyr or earlier have been published since 1990, excluding cases where there are concerns over technical aspects of the measurement of age (see Figure 4.1 for locations of sites). There are thirteen such sites. At three – Lake Mungo, Devils Lair and Ngarrabullgan – occupation at around 40–45 kyr ago has been corroborated by two or more different dating techniques. A variety of materials has been dated to this time interval, including in a few cases organic material which must have been brought to sites by people or produced as a result of human activity (this applies to Gregory River,

Site name	Earliest date (kyr)	Dating techniques	Reference
Allens Cave, WA	40	OSL	Roberts *et al.* (1996)
Buang Merabak, NG	43	^{14}C (AMS)	Leavesley *et al.* (2002)
Carpenters Gap, WA	44	^{14}C (ABOX/AMS)	Fifield *et al.* (2001)
Devils Lair, WA	44.5–48	^{14}C(ABOX/AMS), OSL, U-series, ESR	Turney *et al.* (2001a)
Gregory River 8, Qld	40	^{14}C	Slack *et al.* (2004)
Huon Peninsula, NG	44–51*	U-series, TL	Groube *et al.* (1986); Chappell (2002); O'Connell & Allen (2004)
Lake Mungo, NSW	46–50	OSL	Bowler *et al.* (2003)
Lake Tandou, NSW	41	^{14}C	Balme & Hope (1990)
Malakunanja, NT	45–61	TL	Roberts *et al.* (1990b)
Nauwalabila, NT	53.4–60.3	OSL	Roberts *et al.* (1994)
Ngarrabullgan, Qld	41	^{14}C (ABOX/AMS), OSL	David (2002); David *et al.* (1997)
Riwi, WA	44.5	^{14}C (ABOX/AMS)	Fifield *et al.* (2001)
Yombon, NG	41.5	^{14}C	Pavlides & Gosden (1994)

* *The older date is the youngest of several age determinations provided by Chappell (2002) on the tephra layer which constrains the maximum age of artefacts in this sequence.*

Table 4.1 Archaeological sites from Greater Australia with first-occupation dates of 40 kyr ago or older

Buang Merabak, Lake Tandou and Riwi). This leaves no doubt that people were in Greater Australia by 40 kyr ago, and the evidence for occupation as early as 45 kyr ago is strong (Gillespie 2002; O'Connell & Allen 2004).

Dates on some sites point to still earlier occupation, perhaps extending back as far as 60 kyr, but these earlier dates are controversial. In all of these cases the ages of the deepest artefacts in a sequence were inferred by finding the burial age of sediments lying above and below them and using these to fix upper and lower constraining dates on the artefacts themselves. This can lead to two kinds of problem. First, the age ranges may be very wide. In a few cases (Huon Peninsula, Malakunanja and Devils Lair) they overlap 45 kyr and so do not provide strong grounds for extending the chronology implied by the majority of sites. Second, it is possible that the deepest artefacts in a sequence may have been displaced downwards from their original deposition levels so that they lie in sediments that are too old. Such displacement might have been due to the treading down of artefacts by the people who lived at the site, or disturbance of sediments by burrowing animals. It could have a very large effect on interpretations of age, because a small number of stone tools pushed down by a few centimetres can inflate the estimated first occupation of a site by many thousands of years. The suspicion that this has happened nags at the oldest dates for Devils Lair, Lake Mungo, Malakunanja and Nauwalabila. At all of these sites the densities of artefacts in the deepest levels are very low, a pattern that could indicate displacement of a few objects beyond the true age range of human occupation.

At Lake Mungo there is a high density of artefacts in sediments dated to between 43 and 45 kyr, but below that there is only a sparse scatter of artefacts (Bowler *et al.* 2003). O'Connell and Allen (2004) accepted that the evidence for a human presence at Lake Mungo just after 45 kyr is strong, but they argued that the deeper artefacts were due to downward displacement. At Devils Lair the deepest layers with evidence of people date to approximately 49 kyr, but again they contain only a sparse scatter of artefacts, and they are clearly not in their primary depositional context. Turney *et al.* (2001a) suggested they were washed into the cave from outside and thus represent the first evidence of the presence of people in the region, but O'Connell and Allen (2004) again suggest that they may have been displaced downwards in the sequence. The lowest artefact beds due to primary deposition in the cave are 44.5–48 kyr old. So, both Lake Mungo and Devils Lair can be interpreted as showing initial occupation at about 45 kyr ago.

The dates for Malakunanja and, especially, Nauwalabila have attracted the most controversy. The very early dates for these sites are based on luminescence dating only. Bird *et al.* (2002) tried to corroborate them with advanced radiocarbon methods but could not: the radiocarbon dates on charcoal agreed with the luminescence dates near the surface, but did not increase with depth as they should have done. They attributed this to alteration of carbon at depth during a period when the watertable was raised, and they also suggested that some small particles of charcoal had fallen from upper to lower levels of the site through termite galleries. In both sites the density of artefacts is low in the very deepest levels, and Roberts *et al.* (1990b) commented that there may have been some treading down of artefacts by the occupants of the site. While Bird *et al.* (2002) argued that the effect of this displacement would have been small, other archaeologists worry that the earliest dates for occupation of these sites have been seriously overestimated (O'Connell & Allen 2004).

Roberts *et al.* (1990a) and Bird *et al.* (2002) dealt with the problem of the wide range of constraining dates by constructing age-versus-depth curves for each site and using them to infer the age of the lowest artefacts from their depth in the sequence. This approach provided an estimate of 50 kyr for the level containing the oldest artefact at Malakunanja and 48 kyr for a small peak in artefact density near the base of the artefact sequence at Nauwalabila. The 50 kyr estimate at Malakunanja could probably be reduced by a few thousand years to account for possible downward displacement. So, the data on both sites can be interpreted to give first-occupation dates of around 48 kyr ago.

To summarise: there is no doubt that people were widespread in Australia by 40 kyr ago, and we have solid evidence for occupation by 45 kyr ago. Some careful pruning of the pre-45 kyr age estimates, as outlined above, strengthens the view that people may not have occupied any known sites much before 45 kyr ago, although given the low resolution of dating of material of this age it is impossible to rule out occupation a few thousand years before then. It is possible that people were in Australia around 50 kyr ago or before, and in fact it seems very plausible that small groups of people may have settled in northwestern Australia long before occupation became widespread at around 45 kyr ago. At the moment, however, there is no undisputed evidence of such an early presence.

DISPERSION

Archaeological sites dated to 40 kyr or older are widespread through Greater Australia, with no clear geographic pattern in relation to age (Figure 4.1). The (contested) antiquity of Malankunanja and Nauwalabila hints that the oldest sites might be closest to the point of entry of people, but settlement at Lake Mungo, practically in the opposite corner of the continent, is nearly as old. As far as we can tell from this sparse collection of dates, people spread over Greater Australia within no more than a few thousand years of first setting foot on the continent. There is little evidence of a staged occupation of major environments such as was envisaged by Bowdler (1977). She argued that the first occupants of Australia were a coastal people who would initially have spread around the coast. Occupation of the inland would have been very gradual, depending first on a transition to freshwater economies based on the larger river systems and ultimately on the protracted development of the skills needed for life in dry habitats. It is possible that early occupation was concentrated around the coast and that the evidence of this was drowned under rising seas during the Holocene, but the age-distribution of known sites contradicts Bowdler's model.

A possible exception to rapid occupation is that the most arid regions of central Australia may have been settled rather later than the rest of the continent. The earliest occupation of Allens Cave on the Nullarbor Plain is dated to 40 kyr ago and the two next-oldest sites are Puritjarra (36 kyr, Smith *et al.* 2001) and Kulpi Mara (35 kyr, Thorley 1998). An initial occupation date of 37 kyr for Mandu Mandu Rock-shelter (Morse 1993) suggests that occupation of the arid west coast of Western Australia may have begun at about the same time. However, we should bear in mind that the density of archaeological sites is particularly low in arid Australia and that therefore the chance of finding very old sites, if they exist, is lower than in other parts of Australia. Occupation of Tasmania might also have been delayed by a few thousand years. At present the two oldest sites known from Tasmania are Parmerpar Meethaner (39.5 kyr, Cosgrove 1995) and Warreen Cave (38.5 kyr, O'Connell & Allen 2004). The slightly later occupation of Tasmania could be explained by the fact that before 43 kyr ago Bass Strait was still flooded, and the land connection between northeastern Tasmania and the mainland that opened at that time was initially quite tenuous (Lambeck & Chappell 2001) and may well have been intermittent.

These variations notwithstanding it is remarkable how quickly people settled in and adapted to a broad range of environments in Australia, from tropical savannas in the north to open shrublands and grasslands in the centre and cold steppe in Tasmania. Around 45 kyr ago the climate of Australia was relatively benign. Rainfall was low but not too low, and the fact that it was not as hot as now meant that there was less evaporation and that water availability remained high. Most of the inland was probably semi-arid rather than truly arid, there was still extensive and diverse plant cover, and inland waterbodies such as Lake Eyre were full (see Chapter 5). The speed of settlement of Greater Australia may therefore not be surprising. The next problem is to understand what impact people had on their new home.

THE END OF THE MEGAFAUNA

For decades attempts to find out when the Australian megafauna went extinct were frustrated by technical limits on the dating of megafauna fossils. This changed abruptly with the publication of a large study by Roberts *et al.* (2001a). They dated the remains of 30 taxa of marsupial megafauna as well as two birds and five reptiles in twenty-eight sites from Australia and New Guinea, using OSL to find the burial ages of sediments surrounding remains, supplemented by U-series dating of the crystallisation ages of flowstones lying above or below megafauna-bearing sediments, and checked against radiocarbon ages within the time range of ^{14}C dating. They distinguished sites containing articulated and disarticulated skeletal remains and based their conclusions only on sites with articulated remains. 'Articulation' means that the fossils were found with skeletal elements in correct anatomical association, indicating that they were still in the positions where the animals had lain down to die. Roberts *et al.* applied this criterion because they were unable to directly date bone and so had to infer burial age from the surrounding sediments. A problem with this is that disturbance might have caused the bones to be redeposited in sediments younger than the fossils themselves; this could happen, for example, if they were shifted by water and redeposited, a process that would presumably also result in disarticulation.

The youngest articulated megafauna remains found by Roberts *et al.* (2001a) were 46 kyr old. Statistical modelling of the distribution of ages gave an estimate of 46.4 kyr ago for the extinction event. Two other dating studies published soon afterwards produced results consistent with this extinction date. Pate *et al.* (2002) used AMS ^{14}C dating of charcoal to construct a chronology of megafauna remains at Wet Cave in South Australia and showed that they disappeared just before 45.3 kyr ago, while a range of living species, including some large macropods, persisted throughout the sequence. Pledge *et al.* (2002) used TL to date sediments surrounding articulated *Diprotodon* remains in shallow sediments near Hallett Cove, south of Adelaide, at approximately 55 kyr. These dates are also consistent with studies by Miller *et al.* (1999, 2005), who used a combination of methods to date eggshell of the extinct giant bird *Genyornis newtoni* and the living emu *Dromaius novaehollandiae* in the Lake Eyre region of southcentral Australia. *Genyornis* was common in the area between 140 and 50 kyr ago but had disappeared by 45 kyr ago, while *Dromaius* eggshell was continuously present. The distributions of dates from these studies are shown in Figure 4.2.

These recent studies paint a very clear picture of a brief extinction event that was completed by 45 kyr ago, but not everyone is convinced. The Roberts *et al.* study in particular has attracted some intense criticism. Their paper in *Science* was followed by a rebuttal from Field & Fullagar (2001), an extended exchange in *Australasian Science* (Wroe & Field 2001a, 2001b; Roberts *et al.* 2001b; Gillespie & David 2001) and finally a long critique by Wroe *et al.* (2004c). The two main criticisms were:

1. Roberts *et al.*'s decision to disregard all dates for disarticulated remains was unsound. The critics argue that other criteria can and should be used to test if remains are still in their primary context, and sole reliance on the criterion of articulation allowed Roberts *et al.* to exclude evidence for survival of some species well past 46 kyr ago. Much of this evidence comes from archaeological sites. Mammal remains in archaeological sites are likely to be disarticulated because of handling by people. Ruling out sites with disarticulated remains

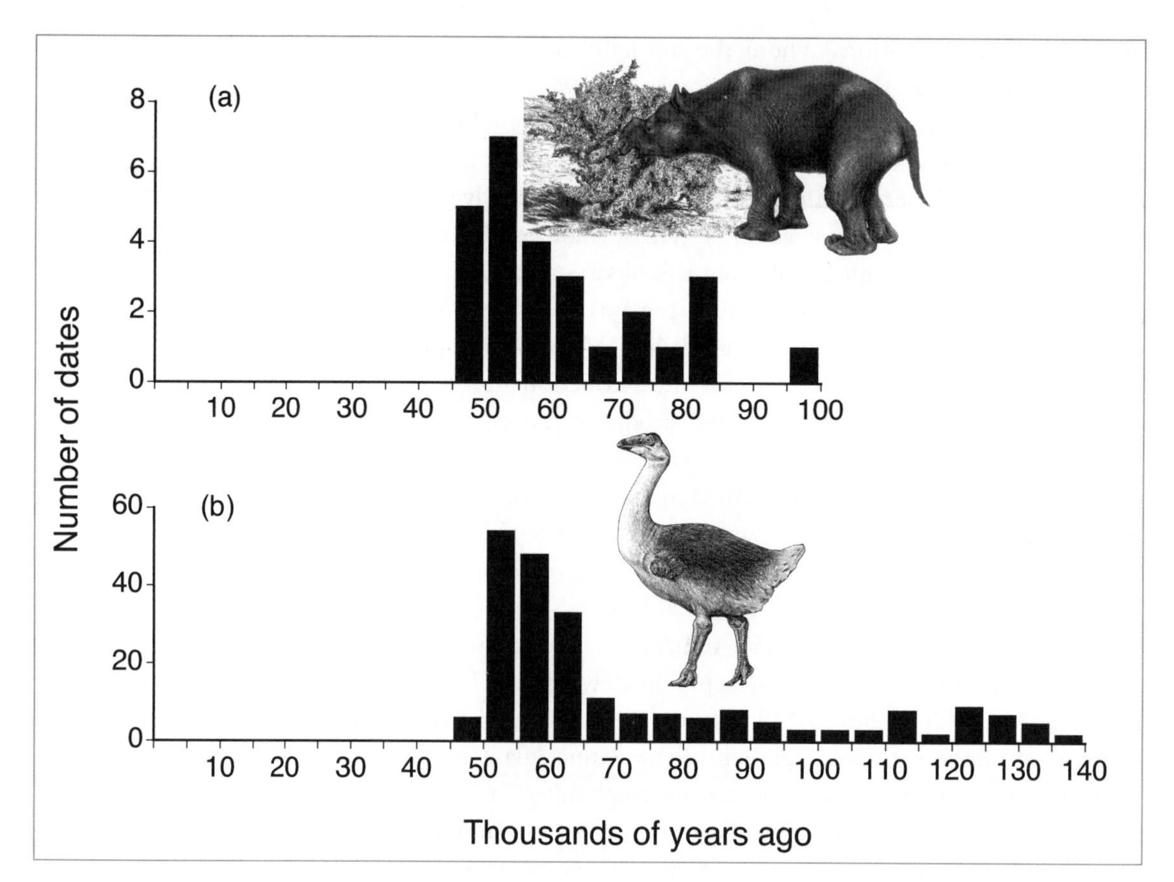

FIGURE 4.2 Age distribution of dated megafauna remains from the last glacial cycle for **(a)** marsupials (data from Roberts *et al.* 2001; Pate *et al.* 2002; Pledge *et al.* 2002) and **(b)** the giant bird *Genyornis newtoni* (data from Miller *et al.* 2005). The studies that produced these data aimed to find the time of dissappearance of species, and therefore focused on megafauna-bearing deposits thought to be relatively young. This explains the tendency for number of dates to peak just before 45 kyr ago.

therefore effectively dismisses archaeological sites, but because almost all such sites in Australia are less than 46 kyr old, ignoring them might close off a rich source of data on late survival of megafauna. By restricting their sample to a small and possibly biased set of remains Roberts *et al.* were 'stacking the odds in favour of finding that extinction occurred earlier than it actually did' (Wroe *et al.* 2004).

2. Even if it were true that no megafauna species survived past 46 kyr ago, Roberts *et al.*'s data do not prove that many of the late Pleistocene megafauna had not gone extinct much earlier. Perhaps 46 kyr marked the quiet conclusion to a long process of gradual depletion of the fauna rather than the dramatic simultaneous extinction of many species.

THE EVIDENCE FOR LATE SURVIVAL

One site has loomed large in the controversy ignited by Roberts *et al.* (2001). Cuddie Springs is a small ephemeral lake between the towns of Walgett and Bourke in semi-arid New South Wales, where megafauna remains are mixed with artefacts through a thickness of sediments implying perhaps eight thousand years of coexistence with people (Field & Dodson 1999). Roberts *et al.* (2001a) obtained ages of less than 36 kyr for these mixed beds, but disregarded them because the remains were disarticulated and also because of variation in dates on individual sand grains which indicated reworking of sediments. In their view the megafauna bones were older than the sediments around them, either because the bones had been redeposited from elsewhere or because the original sediments in the megafauna-bearing layers had been eroded and replaced by younger sediments.

Field and Fullagar (2001) countered with evidence supporting the stratigraphic integrity of the site. First, the bones and artefacts occur in a series of horizons with distinct assemblage characteristics. Second, some megafauna bones in the earliest archaeological levels are in close anatomical position (they may qualify as being 'articulated', and Wroe *et al.* 2004c also complained that Roberts *et al.* had not provided a working definition of this condition). Perhaps most importantly 'the human/megafauna overlap is sealed at its upper and lower limits by consolidated old land surfaces, precluding movement of older or younger material into the horizon after its formation'. Roberts *et al.* (2001b) and Gillespie & David (2001) responded with more evidence of disturbance: some bones are oriented vertically; they show varying degrees of mineralisation; and the proteins they contain are very degraded, suggesting they are older than the sediments in which they now lie. This last point was countered by Coltrain *et al.* (2004), who showed that megafauna bones at Cuddie Springs indeed contain little collagen (which is a pity, as collagen can be dated by radiocarbon) but argued that this is not inconsistent with an age of between 30 and 40 kyr. Most recently Trueman *et al.* (2005) showed that the bones in different strata at Cuddie Springs have distinct elemental signatures, suggesting that they have not been resorted. However, this does not prove that they are of the same age as the sediments around them – it is still possible that the original sediments were eroded and replaced by younger sediments.

Cuddie Springs is significant, and maybe anomalous, in another area of archaeology. Specialised seed-grinding stones are widespread through inland Australia but most are only a few hundred years old (David 2002) – except, it seems, at Cuddie Springs, where grindstones are claimed to be around 30 kyr old (Fullagar & Field 1997) and have been found beneath the consolidated land surface that seals the levels of claimed human/megafauna overlap. David (2002) argued that this feature – essentially a stone pavement – may actually be recent, and he suggested that young sediments containing artefacts may have been redeposited and mixed through older layers with megafauna bones. There is, however, no strong evidence (apart from the location of the offending grindstones) that the pavement is a recent feature, and Field (personal communication) regards this suggestion as extremely unlikely. The debate over Cuddie Springs seems to have reached a stalemate, and will be resolved to general satisfaction only by direct dating of the bones themselves.*

There are several other sites which have been cited as evidence for late survival of megafauna. Of these the strongest claims have been made for Lancefield Swamp in Victoria, originally described by Gillespie *et al.* (1978). The key level in the stratigraphy of this site is a bone bed lying directly on a bed of gravel and between layers of swampy clay. There are very many bones in this bed: Gillespie *et al.* (1978) estimated that it contains about 10 000 individuals of a total of six megafauna species (although the great majority are from *Macropus giganteus/titan*, the giant late Pleistocene form of the living eastern grey kangaroo). A charcoal sample from below the bone bed gave an age (by radiocarbon) of 31 kyr. Two artefacts were found in the bone bed, and more in the overlying clay. Gillespie *et al.* (1978) noted some evidence of movement and sorting of the bones by water, but they regarded this as minor. According to Horton (2000), Lancefield Swamp provides decisive evidence for survival of megafauna many thousands of years after the arrival of people.

This site has now been reinvestigated twice. First, Van Huet (1999), working at a slightly different dig site in the same formation as that studied by Gillespie *et al.* (1978), concluded that the bone bed had been deposited by fast-moving water long after the animals died. The bones show evidence of pre-depositional weathering; they tend to be aligned along a common axis, indicating the direction of water flow; all are disarticulated (and, as Gillespie *et al.* 1978 had noted, there are no complete skulls in the deposit); and the majority are large, suggesting that lighter bones were carried away from the site and only heavy material settled there. Probably the bones were washed into the site at a time when a fast-moving stream flowed through it, in contrast to earlier and later periods when the clays indicate stagnant swampy conditions. Van Huet *et al.* (1998) attempted direct dating of small numbers of *Diprotodon* teeth from the bone bed,

* The latest development in this debate is the publication by Gillespie and Brook of a detailed critique arguing that the stone tools and charcoal in the site were deposited by Holocene floods and mixed with fossil bones of Pleistocene age, and that the stone pavement was constructed during the digging of a well in the 19th century, which caused further mixing of bones, tools and charcoal. This would mean that Cuddie Springs can tell us nothing about events in the last glacial cycle; Gillespie and Brook's interpretation has been rejected in media comment by Field. See Gillespie & Brook (2006).

yielding ages of 46–56 kyr by ESR, 30–55 kyr by AAR, and a minimum of 32 kyr by radiocarbon. All these dates are consistent with extinction by 46 kyr ago, although later survival was not yet completely ruled out.

Next, Dortch (personal communication) re-excavated at the site originally investigated in the 1970s. He found, using the same criteria applied by Van Huet, that the bones were probably in their primary context and had not been redeposited by water. But he dated the sediments in the bone bed by OSL and found that they were between 70 and 80 kyr old. He interpreted the site as a collection of animals that had died around this swampy waterhole during drought, and this is consistent with evidence that the period between 70 and 80 kyr ago was unusually arid (see Figure 5.1). Clearly the small number of artefacts found in the bone bed by Gillespie *et al.* (1978) were intrusive. There is a similar story for Lake

TABLE 4.2 Sites at which survival past 46 kyr ago has been claimed for Australia's extinct megafauna

SITE	LATEST AGE (kyr)	SPECIES[1]	REFERENCE
Cloggs Cave, Vic*	28	*Simosthenurus occidentalis*	Flood (1974)
Cuddie Springs, NSW*	36–30	*Diprotodon optatum, Sthenurus andersoni, Procoptodon* sp., (plus *Genyornis newtoni* and several reptile megafauna)	Trueman *et al.* (2005)
Devils Lair, WA*[2]	38, ~28	*Procoptodon browneorum,* *Protemnodon brehus*	Balme *et al.* (1978)
Kelangurr Cave, NG	16	*Maokopia ronaldi, Protemnodon hopei*	Roberts *et al.* (2001b)
Lake George, NSW	25–31.5	*Procoptodon goliah*	Sanson *et al.* (1980)
Lake Tandou, NSW*	25	*Procoptodon* sp.	Hope *et al.* (1983)
Lime Springs, NSW*	23.5	*Diprotodon* sp., *Protemnodon* sp., *Procoptodon* sp., *Sthenurus* sp.	Gorecki *et al.* (1984)
Lancefield Swamp, Vic*	31	*Protemnodon anak, P. brehus, Simosthenurus occidentalis, Diprotodon* sp.	Gillespie *et al.* (1978)
Main Cave, Tas*	13	*Simosthenurus occidentalis, Zaglossus* sp.	Goede & Bada (1985)
Mooki River, NSW	42	*Thylacoleo carnifex, Palorchestes azael, Phascolonus gigas, Diprotodon optatum*	Roberts *et al.* (2001b)
Nombe Rock-shelter, NG*	29–17	*Protemnodon nombe*	Flannery *et al.* (1983)
	17–11.5	*P. tumbuna*	
Seton Rock-shelter, SA*	19	*Procoptodon gilli*	Hope *et al.* (1977)
Spring Creek, Vic*[3]	38	*Palorchestes azael*	White & Flannery (1995)
Tight Entrance Cave	33	*Simosthenurus occidentalis*	Roberts *et al.* (2001b)

* *Sites with artefacts*

1 *Not including species considered to be large late Pleistocene forms of living species (e.g.* Macropus titan*)*

2 *Refers only to layers above those also dated by Roberts et al. (2001b)*

3 *Average for two differing dates on the same sample of bone*

Menindee in western New South Wales, where Tedford (1955) and Tindale (1964) found megafauna deposits associated with archaeological material thought to be between 23 and 31 kyr old. Cupper (personal communication) has redated the site and found that the archaeological and megafauna deposits are distinct and that the bones date to between 46 and 55 kyr ago.

Table 4.2 lists other sites at which extinct megafauna have been dated to less than 46 kyr (see Figure 4.1 for locations). In all cases the remains are disarticulated, and eleven sites contain evidence of human occupation. There is stratigraphic overlap of artefacts and megafauna remains at five sites: Devils Lair, Lime Springs and Nombe Rock-shelter, as well as Cuddie Springs and Lancefield. At Seton Rock-shelter and Lake Tandou stratigraphic overlap is a possibility but is marginal at best; overlap has also been reported at Keilor in Victoria (Duncan 2001) and Lembudu Cave on the Aru Islands (O'Connor *et al.* 2002), but without precise dates.

Are the young dates from any of these sites reliable? Dates for several sites must be considered doubtful on the basis of their investigators' own statements of the possibility that megafauna remains in recent layers had been redeposited from older strata. This applies to Lake Tandou, Main Cave, Devils Lair and Kelangurr Cave. In other cases where authors were confident of the stratigraphic integrity of their sites, reworking should still be admitted as a possibility, because megafauna remains are in very small quantities (Cloggs Cave, Seton Rock-shelter, Keilor, Lembudu Cave) or are highly fragmented (Lime Springs). The original description of Nombe Rock-shelter regarded the site as being undisturbed, but according to Flannery *et al.* (1996) there is now evidence of some reworking (people may have dug holes in the cave floor and disturbed the sediments in recent times).

Direct dating of megafauna bone has been attempted at a few of these sites. Görecki *et al.* (1984) claimed overlap of humans and megafauna at about 24 kyr ago at Lime Springs, and Wright (1986) followed this with claims that megafauna survived into the Holocene at this site and the nearby Tambar Springs and Trinkey sites. Fethney *et al.* (1987, cited in Flood 1999) measured ages in the range of 23–40 kyr ago by U-series dating of *Diprotodon* teeth. This would seem to rule out Holocene survival, but Fethney *et al.* (1987) noted serious anomalies in their results and concluded that the true ages of the remains were unresolved. Goede & Bada (1985) used ESR dating of bone to test earlier inferences of overlap of humans and megafauna in several cave sites in southwest Tasmania (Goede & Murray 1977; Goede *et al.* 1978). They found that most megafauna-bearing deposits were much older than previously thought, but they described a previously unreported bone deposit containing some megafauna at Main Cave and dated to 13 kyr ago (although it seems the megafauna bones themselves were not dated, and the possibility of reworking remains). White & Flannery (1995) obtained dates of 25.4 and 36.5 kyr (by radiocarbon) on a *Palorchestes* bone from Spring Creek in Victoria, a site for which an earlier claim of megafauna/human overlap at about 20 kyr ago had been made. The two dates were from different portions of the same bone dated by the same method in different laboratories, and the large discrepancy suggests a serious problem with the accuracy of the dating. At Rocky River on Kangaroo Island, Forbes *et al.* (2004) dated soil organic matter associated with remains of *Diprotodon*, *Zygomaturus* and *Sthenurus* species to between 18.5 and 21 kyr old. However, radiocarbon dates on soil organics are unreliable because that material is readily contaminated by younger carbon. Direct dating of bone at these sites is underway, and suggests that they range in age from 45 to 110 kyr old (Wells, Grün, personal communication).

To summarise: there are reasons to suspect reworking of sediments at many of the sites listed in Table 4.2. In most cases where claims of late survival of megafauna have been re-evaluated, they have turned out to have been erroneous. The trend in the results published most recently is for terminal dates to be concentrated at 45 kyr ago or a little before. There are now very few sites with substantial megafauna remains associated with dates younger than 45 kyr: Tight Entrance Cave (Prideaux *et al.* 2000 and Prideaux personal communication) and possibly Nombe Rock-shelter may fall into this category. Late survival of megafauna at Cuddie Springs is strongly defended by the site's excavators, but others are deeply sceptical.

EXTINCT MEGAFAUNA AS 'DREAMTIME ANIMALS' There is another kind of evidence that might reveal late survival of megafauna. If the extinctions were completed after the arrival of people, then we might find signs in the human cultural record of contact with megafauna. These could take the form of rock paintings of the extinct animals, legends in which they play a part, markings on bones, or the incorporation of megafauna remains in artefacts. A long period of interaction between people and megafauna might have produced a large body of such evidence.

The rock art of northern Australia is the oldest on the continent and among the most ancient in the world. Some paintings that can still be seen in the rock galleries of Arnhem Land and the Kimberley may have been made soon after people first arrived in Australia (Morwood 2002). Strange animals do appear in some of the earliest of these paintings. Murray & Chaloupka (1984) note that well-executed echidna paintings are common in early rock art styles of Arnhem Land. They seem to show two different types of echidna: one like the living short-beaked echidna, and another with a long tapering head and a humped back that looks very much like a long-beaked echidna *Zaglossus* (Plate 9). Long-beaked echidnas still survive in New Guinea, but the Australian *Zaglossus* disappeared with the other megafauna.

Many Arnhem Land paintings contain accurate depictions of thylacines, and two of them show animals that resemble thylacines except that they have distinct tail-tufts, tail butts that are clearly demarcated from the body, broad paws, limb proportions that match those of the marsupial lion *Thylacoleo*, and no stripes. (Murray and Chalouka comment of one of these pictures that the animal 'is shown in an "actionless" or dead posture, with its limbs hanging motionless and its feet pointed downwards as though the actual specimen was laid out for inspection while it was being drawn'). Other Arnhem Land paintings show kangaroo tracks with only a single large toe that might represent sthenurine kangaroos, and there is a very obscure painting of a strange elongate mammal that might just be a *Palorchestes*. In the Kimberley there are early paintings of a large-bodied, short-legged marsupial interpretable as a diprotodon, more single-toed kangaroos, and other broadly *Thylacoleo*-like animals (Akerman 1998; Walsh 2000).

The idea that early Aborigines saw megafauna species and rendered them in artworks that have survived more than 40 kyr is wonderful, but Lewis (1986) argues that we know too little about the conventions followed by the artists of the time to interpret these depictions of strange animals. He also suggested that some of the unusual body forms described by Murray and Chaloupka were due to superimposition of different paintings of living species. Despite these misgivings some of the paintings they described are compelling, especially those of the long-beaked echidna.

There are some oral traditions that might refer to extinct megafauna, but it is even harder to be certain that these were originally inspired by experience of real animals. For example, the Adnyamathanha people of the Flinders Ranges tell stories of a legendary mammal known as Yamuti (Tunbridge 1991). Yamuti was bigger than any living native mammal; he was dangerous, and ate people; he was unable to raise his head to look up towards the sky, so that children were told they could escape him by climbing a tree; and he sometimes took the form of a giant kangaroo. There are no anatomical details in these stories that could unambiguously identify Yamuti with any particular megafauna species, but it is tempting to see this legendary beast as a kind of chimera, combining various features of real animals in a single creature. The stories also reveal a particularly strong

PLATE 9 Rock painting from Arnhem Land depicting an extinct long-beaked echidna *Zaglossus* sp. (photograph by George Chaloupka). The inset shows a sketch of a living *Zaglossus* from New Guinea. Note the paint strokes representing long spines over the animal's hindquarters.

PLATE 10 Spinifex-dominated landscape near the Bungle Bungle Range in the Kimberley (photograph by Euan Ritchie). Spinifex became more widespread in this region about 30 kyr ago, and in the last few thousand years has probably increased further as a result of extensive landscape burning by people.

PLATE 11 All the largest living marsupials are grazers, but the extinct megafauna included many very large specialised browsers. (photograph of antilopine wallaroos in Kimberley grassland by Jen Martin)

PLATE 12 Anti-browser plant architecture in juvenile scrub leopardwood *Flindersia dissosperma*. Note the wide angle of branching, stiff and spiky twig tips, and small leaves widely separated along branchlets. (photograph by Jon Luly)

element of fear of a large carnivore. Adnyamathanha people believed that *Diprotodon* remains from Lake Callabonna actually belonged to Yamuti.

Perhaps the most famous of the megafauna traditions is one held by the Tjapwurong people of western Victoria concerning a giant emu called the mihirung, said to have lived at a time when there were active volcanoes in the region (Murray & Vickers-Rich 2004). This legend could conceivably be based on *Genyornis newtoni*, the giant bird that went extinct along with the marsupial megafauna. There is even evidence from sediments in the bed of Lake Wangoom in western Victoria that there was a volcanic eruption about 45 kyr ago (Harle *et al.* 2002). At the Mootwingee rock art site in western New South Wales there is a rock panel with engravings of emu footprints beside a much larger set of engraved prints consistent with *Genyornis* (Murray & Vickers-Rich 2004).

Archer *et al.* (1980) described incisions, breakage patterns and charring on megafauna bones from Mammoth Cave in the southwest of Western Australia that might have been produced by people. A *Sthenurus* tibia contained a notch consistent with a stone tool, and many long-bones were broken as if pounded by stone hammers (perhaps to remove marrow). Some bones were charred in a way that implied they had been laid over a cooking fire. These bones appeared to have been deposited within a short period of time, and the deposit was dated by radiocarbon at around 42 kyr old (although given the limitations on the radiocarbon method at the time, this should be regarded as a minimum estimate of age). The accumulation of bones contained no stone tools, and Archer *et al.* suggest that it represented the remains of animals that had been used by people then thrown from the surface down vents into the cave. Vanderwal & Fullagar (1989) described a series of parallel grooves on a diprotodon tooth from Spring Creek that they judged to have been made by longitudinal cutting strokes, possibly from a flint flake, rather than by vertical or crushing blows from a predator Most intriguing of all are two charms from the Kimberley region, one a *Zygomaturus* premolar mounted in gum, the other a small collection of teeth from *Procoptodon brownoreum* held in an emu skin wallet. Akerman (1973) assumed these teeth had been collected relatively recently from an (unknown) exposed fossil deposit, but Long *et al.* (2002) allowed the possibility that they were actually collected from live animals and passed down as heirlooms.

These examples are tantalising, but it is remarkable how thin this evidence is. There may have been little real hope that unmistakable images of megafauna could have survived 45 kyr or so in rock paintings, and still less that definitive descriptions of them might have lasted in oral traditions. But, we might have expected more human-marked megafauna bones. Maybe the problem is that poor preservation of bone from the earliest archaeological sites means that much material evidence of interaction between people and megafauna has been lost. In archaeological sites that extend beyond 40 kyr in age the oldest levels contain little or no bone. For example, at Riwi, Carpenters Gap and Buang Merabak the density of bone shows a general decline with increasing age (Balme 2000; Leavesley & Allen 1998; Wallis 2001); there is little bone, and most of it is fragmented and burnt, in the deepest levels at Lake Mungo; and there is no bone at all in the older levels of Nauwalabila and Malakunanja because the soil is too acidic for its preservation (Jones & Johnson 1985).

Poor preservation might explain the lack of megafauna bones marked by people, but it is hard to escape the conclusion that the megafauna disappeared quite soon after the arrival of people, or that if they persisted for long they were very rare and most people did not come into contact with them. This would also explain why there are so few archaeological sites with evidence of overlap between humans and megafauna. Over 160 Pleistocene-age archaeological sites are known from Australia and New Guinea (Smith & Sharp 1993), but only a small handful have any evidence of humans and megafauna in association and, as we just saw, there are doubts hanging over all of these. The lack of evidence for overlap of humans and megafauna is particularly striking in southwest Tasmania. Here, sites with evidence of human occupation during the Pleistocene contain very large accumulations of well-preserved and readily identifiable mammal bones – Cosgrove & Allen (2001) review a collection of 637 000 bones from more than a dozen sites – none of which belongs to megafauna species.

A BANG OR A WHIMPER?

Not all the extinct mammals listed in Table 2.1 are known to have been still alive at any time in the last glacial cycle. Of the total, there is evidence that 35 made it into the last glacial (although the quality of this evidence varies). Does this mean that the others were already extinct? Not necessarily, and here we run into a problem known to palaeontologists as the Signor-Lipps Effect. The problem is that because fossils are rare it is very difficult to use the fossil record to pinpoint the extinction date for a species. If we have only a few fossils for a given species the youngest fossil might pre-date the animal's actual extinction by many thousands of years, even by hundreds of thousands of years. So, even if all the species in Table 2.1 made it into the last glacial cycle and went extinct at precisely the same time, the terminal dates for individual species revealed by the fossil record would be spread through some long time interval leading up to that true mass-extinction date. If we made the mistake of assuming that the youngest fossil for each species represented its extinction date we would wrongly conclude that the megafauna had declined gradually rather than disappearing all at once. The age of the youngest fossil of a species tells us only that the species went extinct some time *after* that date (unless the individual preserved as that fossil was really the last of its kind).

The distortion caused by the Signor-Lipps Effect is worst for species that left few fossils, but for species that commonly fossilised we have a better chance of reading a true extinction date from the fossil record. So, if there really was a mass extinction at 46 kyr ago we should expect the fossil evidence for survival close to that date to be strongest for the species with a more abundant fossil record. This is tested in Figure 4.3 for extinct mammal species from Pleistocene sites in the Darling Downs region (Molnar & Kurz 1997) and the southeastern region of South Australia (Reed & Bourne 2000). In each case there are lists of species that have been recorded in each of a large number of sites spread over the region. I used the proportion of sites at which a species has been found in each region as an index of its fossil abundance, and I averaged these values across both regions. I then checked to see if there was a relationship between abundance of fossils and the probability that a species has been

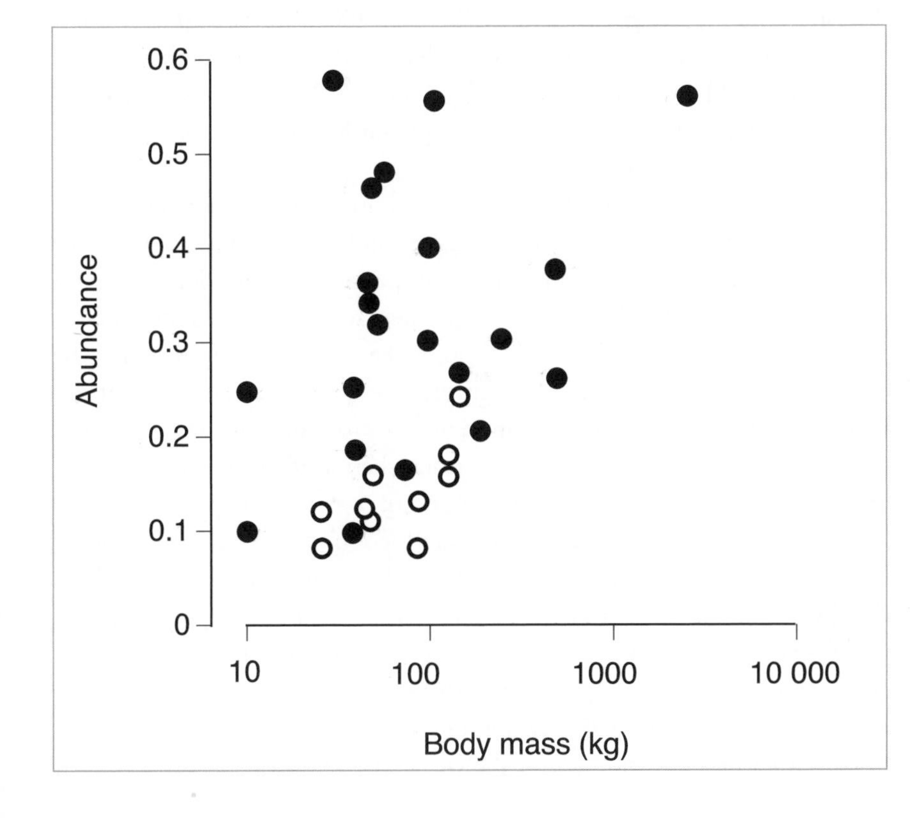

FIGURE 4.3 The abundance of megafauna species in the Pleistocene fossil record of the Darling Downs and southeastern South Australia, distinguished by whether these species are known to have survived into the last glacial cycle (●) or not (○).
Notes: Abundance is measured as proportion (square root transformed) of Pleistocene sites in the region at which each species was recorded (data from Molnar & Kurz 1997, Reed and Bourne 2000). Mean abundance of the two classes of species differs significantly ($t = 4.02$; $p < .001$).

recorded, somewhere in Australia, in the last glacial cycle. The figure shows that this relationship is positive, as one would expect if there really was a synchronised extinction in the last glacial cycle that is partly obscured by a Signor-Lipps Effect.

Similarly Prideaux (2004) noted that there are only 11 species of sthenurines with relatively complete fossils represented by articulated remains and, of these, 8 occur in deposits dated to the late Pleistocene; deposits in which the other three occur await dating. Prideaux's review of the fossil record of the sthenurine kangaroos, the most species-rich group of Australia's late Pleistocene large mammals, revealed no evidence of decline before the last glacial cycle. If anything, species richness increased steadily through the Pleistocene until the last glacial cycle, when the group disappeared completely. Prideaux (personal communication) provided a thorough census of a middle Pleistocene mammal community at Cathedral Cave, in a region that also has well-sampled late Pleistocene deposits. The fauna included 13 species that were extinct by the Holocene (two of which, the large devil *Sarcophilus laniarius* and large koala *Phascolarctos stirtoni*, may be ancestors of living species); of these only one, *Simosthenurus pales*, has not yet been recorded from the last glacial cycle. Dating of other sites in the Naracoorte Caves over the last 500 kyr show little or no change in faunal diversity until the last glacial cycle (Grün *et al.* 2001; Moriarty *et al.* 2000).

A sensible conclusion from this evidence is that many, perhaps all, of the species of Pleistocene mammals for which we have no fossils from the last glacial cycle were still alive at the beginning of that cycle, but had gone extinct by its end.

The Signor-Lipps Effect has two other implications for our understanding of late Pleistocene palaeontology and archaeology. First, it also works in reverse; that is, originations as well as extinctions will seem to be spread out in the fossil record even if they were actually simultaneous (Sepkoski 1998 suggested that this should be called the Lipps-Signor or perhaps the Sppil-Rongis effect). This gives us a reason to suspect that the first appearances of people in different parts of Australia were actually closer together in time, and a little earlier, than a literal interpretation of the opening dates would suggest. Second, if a common species is made rare it might survive at reduced numbers for some long period but never be registered in the fossil record. Therefore the sudden disappearance from the fossil record of a hitherto well-sampled species could represent either a well-dated extinction event or an abrupt decline to rarity. So, if Roberts *et al.* were correct in their identification of 46 kyr ago as a significant boundary in the history of the Pleistocene megafauna, the event dated might either be the extinction of the fauna or its collapse to rarity. Indeed it could well be a combination of the two: some species going extinct at that time, others persisting for a while at low numbers that made them palaeontologically invisible before eventually dying out.

The evidence at present supports the following statements about the chronology of the megafauna extinctions. Before the last glacial cycle Australia's diverse megafauna assemblage was stable, but something then destroyed it. Many species in this assemblage, perhaps all of them, persisted through the first 70 kyr of this last cycle but were gone by 45 kyr ago. The evidence for continued existence of megafauna for a substantial period after 45 kyr is thin, and may stand or fall on the interpretation of a single controversial site (Cuddie Springs). There are hints that the first people in Australia encountered megafauna, but they are faint hints, and this supports the view that the megafauna either went extinct, or became extremely rare, very soon after the arrival of people.

5

The changing environment of the Pleistocene

AS FAR AS we can be confident over the dates, the period between 40 and 50 kyr was when people occupied Australia, and when the megafauna disappeared. This chapter puts that period into the context of changes in climate, vegetation and fire through the whole of the last glacial cycle, and compares the last glacial cycle with the preceding ones. It shows how the habitats of megafauna species changed through the last glacial cycle, and answers two critical questions. Was the arrival of people followed by increased burning and vegetation change? And, were environmental changes in the last cycle more extreme than in earlier ones?

ICE AGE CLIMATES

TEMPERATURE AND CO_2

For the last two million years or so the climate of the world has been dominated by cycles in temperature, known as the glacial cycles because they caused successive expansions and contractions of ice sheets in the northern hemisphere. In Australia there was little glaciation other than in the southeastern highlands and parts of Tasmania, but the 'ice ages' brought cool and dry conditions to the continent. The major features of these cycles are shown in Figure 5.1, which identifies several major points. First, the long-term decline in average temperature came to an end about 700 kyr ago (Figure 5.1a). Second, the range of temperature cycling has widened, with an increase in amplitude around 700 kyr ago due to lower temperatures at glacial maxima, and another at 450 kyr ago due to warmer spells between glacial cycles. Also, there was an increase in the cycle length about 700 kyr ago, by extension of the cold phases from about 40 to 100 kyr. Third, the last four cycles settled into a very consistent pattern: an abrupt drop in temperature at the beginning of each cycle was followed by a

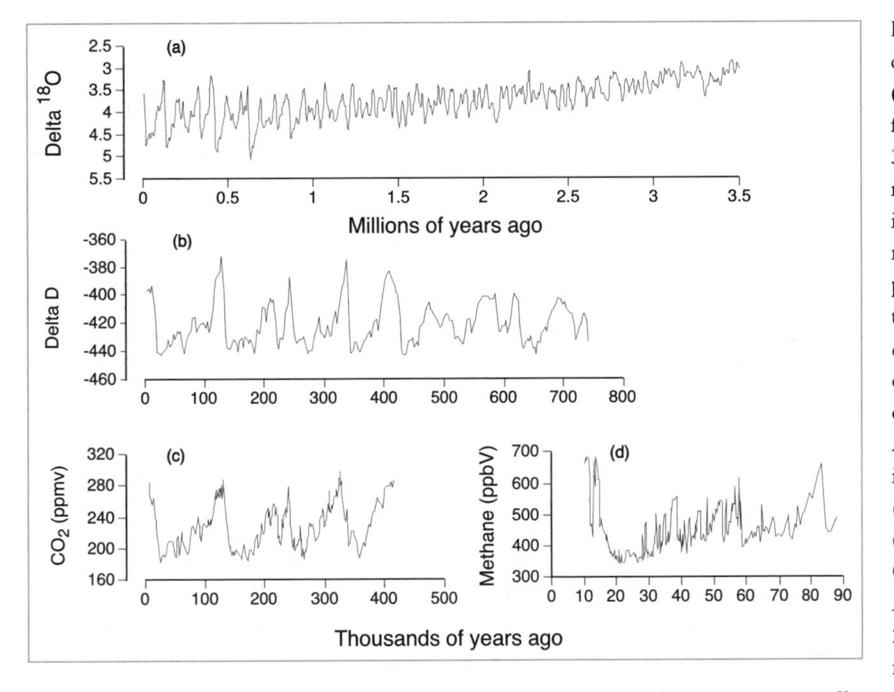

FIGURE 5.1 Pleistocene climate cycles: **(a)** temperature fluctuations over the last 3.5 million years, as measured by oxygen isotope ratios in fossil marine organisms; high points indicate high temperatures (from Mix *et al.* 1995); **(b)** the last eight glacial cycles, from deuterium ratios in Antarctic ice; high points indicate high temperatures (from EPICA 2004); **(c)** cycles in atmospheric CO_2 concentrations in Antarctic ice (EPICA 2004); **(d)** climate fluctuations within the last cycle, indicated by methane concentrations in Antarctic ice (Blunier & Brook 2001).

period of about 100 kyr when temperatures fluctuated over a generally downwards path before dropping into an extreme cold phase close to the end of the cycle, after which there was a sharp rebound to warm conditions (Figure 5.1b). The warm intervals between glacial periods – the interglacials – have usually lasted only a few thousand years (the one in which we are now living has already gone on for an unusually long time at 10 kyr).

Within the last glacial cycle there was a period of moderate temperature between about 105 and 75 kyr and again between 60 and 40 kyr, and a mid-cycle cool spell around 70 kyr ago. The coldest stage of the last cycle – the Last Glacial Maximum (LGM) – began about 35 kyr ago, reached its extreme just before 20 kyr ago, and ended at about 18 kyr ago. Mean atmospheric temperatures over southern Australia shifted through a range of about 10 degrees C from interglacials to glacial maxima (Miller *et al.* 1997; Petit *et al.* 1999); the temperature decline was less pronounced in the tropics than at higher latitudes (Barrows & Juggins 2005).

These Pleistocene fluctuations in temperature have been accompanied by cycles of atmospheric CO_2, from high CO_2 in warm conditions to low CO_2 in cool conditions. What causes these CO_2 cycles is not well understood (Francois 2004), but they are dramatic: from peak levels during glacials, CO_2 concentrations declined by more than 35 per cent to their lowest levels at glacial maxima (Figure 5.1c). The amplitude of CO_2 cycling increased after 430 kyr ago, as temperature cycling also became more dramatic (Siegenthaler *et al.* 2005).

WATER

Because of lower atmospheric and sea surface temperatures and falling sea levels, rainfall declined during glacial cycles. It seems that under glacial conditions vast areas of inland Australia turned into cool deserts as arid environments spread from the centre towards the margins of the continent (Nanson *et al.* 1992). At least that is implied by the evidence for increased sand dune activity over much of the inland from about 30 until 20 kyr ago (Munyikwa 2005), extending even into places like the Blue Mountains west of Sydney that now have relatively high rainfall and dense vegetation cover. At the same time there were sharp increases in the volume of continental dust blown offshore to settle in marine sediments to the southeast and northwest of Australia (Figure 5.2). The extreme conditions – huge dust storms over a denuded landscape – indicated by high rates of dust export were typically limited to short periods at the end of each glacial cycle (except for the second last cycle, which also had an unusually prolonged cold phase, see Figure 5.1). There is some evidence that dust export became significant only within the last 400 kyr. This is suggested by the million-year E39.75 sediment core from the Tasman Sea in which dust concentrations are low and constant until an increase at 380 kyr, a point that Hesse (1994) suggested 'marks the transition to true aridity' in Australia.

The history of dune building and dust export identifies the period from 30 to 20 kyr ago as a time of sparse vegetation cover over much of the present-day inland of Australia. There was an earlier episode of aridity from about 70 to 50 kyr ago that coincided with a dip in atmospheric temperature (Figure 5.1b). However, between 50 and 35 kyr ago the climate appears to have been quite mild and moist, and some of the great inland waterbodies such as Lake Eyre held permanent water (Munyikwa 2005).

Although there was a general increase in aridity towards the end of the glacial cycle, the effects of this on water availability were very different in northern and in southern Australia (Hesse *et al.* 2004; Kershaw 1995). In the north there was a widespread drying of lakes and rivers, which by 20 kyr ago was nearly complete (Bowler *et al.* 2001). Rainfall over northern Australia is dominated by the summer monsoon, which waned under glacial climates so that the aridity that now prevails in the dry season of northern Australia was unbroken. The history of Lake Eyre also shows this. Although it is far to the south of the tropics, Lake Eyre gets its water from rivers that rise in northwestern Queensland. Because an extreme monsoon event is needed to drive a flood all the way down these rivers, the water level in Lake Eyre is a sensitive indicator of the monsoon's changing strength. Around 125 kyr ago Lake Eyre was a deep,

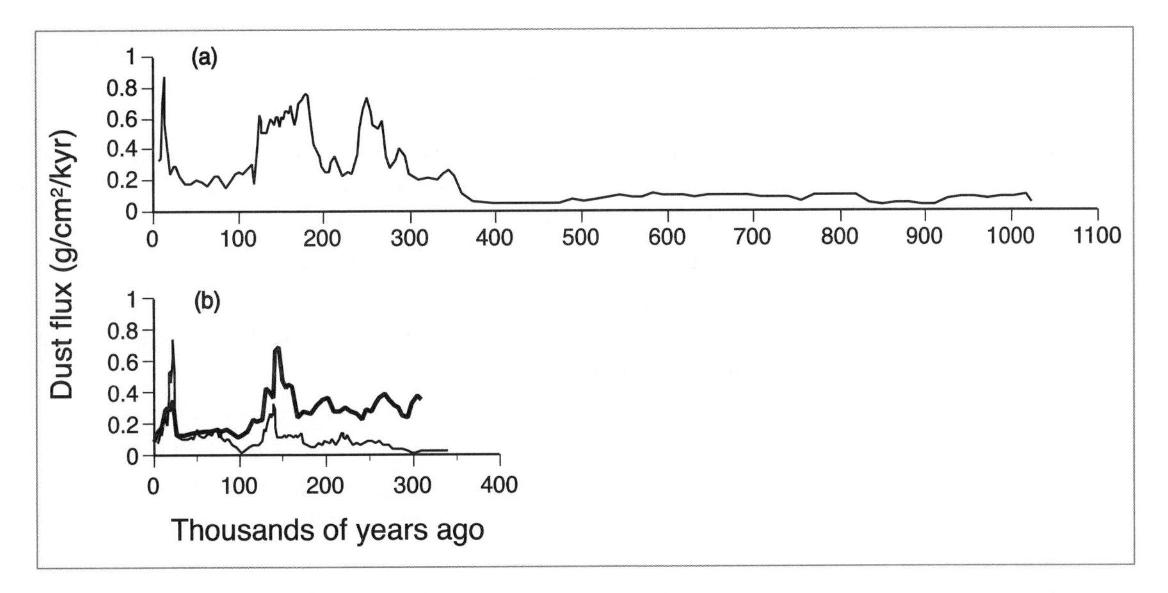

FIGURE 5.2 Records of export of continental dust from the Australian region: **(a)** a million-year record of dust in a marine sediment core (E39.75) from the Tasman Sea; **(b)** two shorter records from marine sediment cores in the Indian Ocean off northwestern Australia (heavy line, core SO-14-08-05), and the Tasman Sea (light line, core E26.1). Redrawn from Hesse & McTainsh (2003) and Hesse *et al.* (2004).

permanent 'mega-lake' with an area of nearly 35 000 square kilometres. The whole Lake Eyre–Lake Frome system held 430 cubic kilometres of water compared with a maximum filling in historical times of only 30 cubic kilometres (DeVogel *et al.* 2004). The lake went through several cycles of drying and refilling between 125 and 30 kyr, but peak water levels were lower on each occasion and it was continuously dry from 30 to 12 kyr, when the monsoon appears to have shut down completely over northern Australia (Magee *et al.* 1995).

By contrast, in some parts of southern Australia there was more water during the LGM than before or after it. Lake George, just north of Canberra, held much more water at the LGM than it has in historic times, and flows in the Murrumbidgee River system may have been four times higher than recent levels (Hesse *et al.* 2004). These differences suggest that the weather patterns that bring winter rainfall to southern Australia held on more strongly through the LGM than did the northern monsoon, but three other factors were probably also important:

1. The development of glaciers meant that streams originating in the southeastern highlands were regularly supplied with water from seasonal melting of ice and snow. This seems to have stabilised water levels in river systems like the Murrumbidgee. The glacial influence also extended to some lakes in the semi-arid inland. An increase in the water level of Lake Mungo at about 60 kyr ago coincided with a major glacial advance on the Kosciuszko massif, and a brief rise in lake levels at 30 kyr ago followed the beginning of another glacial advance at 32 kyr ago (Barrows *et al.* 2001;

Bowler *et al.* 2003). Water levels then dropped, but Lake Mungo continued to hold some water until about 20 kyr ago (Bowler 1998; Bowler *et al.* 2003).

2. Lower temperatures in the south would have reduced evaporation, so the rain that did fall may have been more effective in recharging ground water and delivering water to streams and lakes. This can explain the fact that in South Australian caves there were high rates of deposition of calcite – an indication of high levels of ground water – during some of the coldest parts of the last glacial cycle (Ayliffe *et al.* 1998).
3. The combination of low evaporation with a (probably) much less variable rainfall regime created the conditions for the formation of persistent shallow wetlands in some places that were dry beforehand and are dry now. Williams *et al.* (2001) described sediments of LGM age in valleys of the Flinders Ranges that are fine-textured and well stratified and that contain the remains of snails, diatoms and swamp grasses. These indicate slow-moving streams and perennial wetlands in places where, in our time, water retention is low because watercourses are scoured by flash floods after intense rain.

CLIMATE VARIABILITY

Figure 5.1d shows a more finely resolved picture of climate fluctuations through much of the last glacial cycle, as revealed by methane concentrations through the Byrd ice core from Antarctica (Blunier & Brook 2001). High methane indicates relatively warm and wet global climates (the methane is thought to have been produced in tropical wetlands, and high values reflect expansion of these habitats). This shows clearly something that is hinted at in many other climate records: many short-term fluctuations, on time scales of about one thousand years. These fluctuations were especially dramatic in the middle part of the last cycle but they settled down in the approach to the LGM, and from 28 to just after 20 kyr ago the climate was very stable. There is other evidence that the climate became more stable at the LGM. Sediments of LGM age from the sea floor in the Gulf of Carpentaria, which was once an inland lake, are finely structured and were obviously little-disturbed by intense storms (De Deckker 2001), and detailed reconstructions of sea surface temperatures around the Australian coast show significantly less seasonal temperature variation than at present (Barrows & Juggins 2005).

Nowadays, climate variability in Australia is dominated by the El Niño-Southern Oscillation (ENSO) system. ENSO is the pendulum swing of ocean temperature and atmospheric circulation over the Pacific that brings alternating deep droughts and wet conditions to much of Australia on a cycle of about five years. El Niño events also have dramatic effects on weather elsewhere on the globe. The El Niño of 1788–1795, which dried up the main source of drinking water for the European settlement at Port Jackson in New South Wales, also caused crop failures and general misery in France that may have helped to trigger the French Revolution (Grove 2005), while the 1997–1998 El Niño caused disastrous forest fires in Indonesia.

Climate variability on ENSO timescales can be read from chemical signatures in fossil corals, and these show that ENSO has been operating for at least the last 150 kyr (Tudhope *et al.* 2001). Its strength has varied, however, and there are indications of two major controls on this. First, ENSO variability tends to be greatest during the warm conditions of interglacials and is damped down during glacial periods. Second, climate models predict a 22 kyr cycle in ENSO activity linked to precession – the regular wobble in the orientation of the earth relative to the sun that affects the distribution of solar radiation over the equator (Clement *et al.* 1999). The interaction of these two cycles results in complex variation in ENSO strength (Tudhope *et al.* 2001). Both cycles have fallen together within the last three or four thousand years, which have seen the most severe ENSO cycling of any time in the last 150 kyr. ENSO was also quite strong at the last interglacial, around 120 kyr ago, but was at moderate or low strength for much of the last glacial cycle, and was very weak 6.5 kyr ago.

How ancient is the ENSO system? Kershaw *et al.* (2000, 2003) made a case that there was a rise in ocean temperatures in the western Pacific about 400 kyr ago, which may mark the beginning of ENSO. The ultimate cause of this warming might have been the northward movement of Australia, which gradually restricted the westward flow of Pacific water into the Indian Ocean and trapped a pool of warm water in the western Pacific. This West Pacific Warm Pool is responsible for the temperature gradient across the Pacific, and its formation 400 kyr ago may have been the push that set ENSO in motion.

VEGETATION AND CLIMATE CHANGE

Falling temperatures and rainfall through the glacial cycles favoured drier and more open vegetation types, and moisture-loving plant communities like rainforest contracted. Low atmospheric CO_2 must also have had very significant effects on vegetation. Because plants fix carbon from CO_2 by photosynthesis, they would have been starved of carbon under low-CO_2 glacial conditions. Plants growing under the shade of other plants would have suffered from the combined effects on photosynthesis of low light and low CO_2. A general shortage of carbon would have restricted the capacity of woody plants to engineer structural tissue, so trees and shrubs were probably smaller. Plants with the C3 photosynthetic pathway (temperate grasses, and most shrubs and trees) are more severely affected by such conditions than C4 plants (mainly tropical grasses) because C3 photosynthesis is more demanding of CO_2 Therefore, low CO_2 alone should have produced a more open and structurally simple vegetation, with a radically changed species composition dominated by C4 plants (Ehleringer & Monson 1993).

ENVIRONMENTAL HISTORIES WRITTEN IN POLLEN

The imprint of past vegetation is preserved in pollen buried with sediments in lake beds and on the ocean floor, and the identification and counting of pollen in sediment cores has been critical in reconstructing past vegetation patterns in detail. The locations of those cores that encompass the last glacial cycle – and in some cases that reach back into preceding cycles – are shown in Figure 5.3, and their major features are described below.

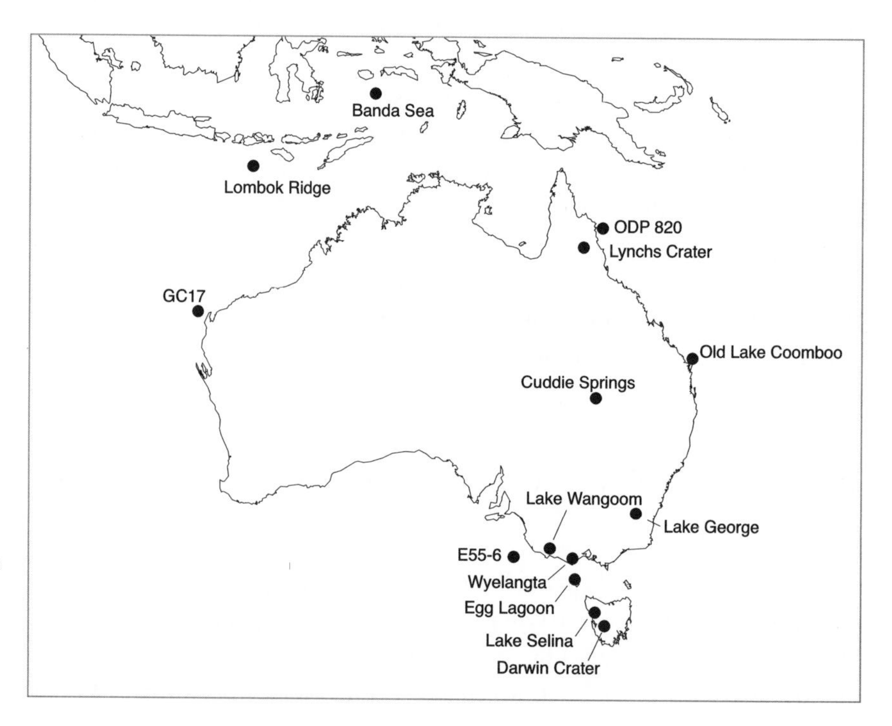

FIGURE 5.3 Map of Australia and islands to the north, showing the locations of vegetation and charcoal records discussed in the text.

THE MOIST TROPICS The most thoroughly studied and influential of these cores are from Lynchs Crater, a small volcanic lake on the Atherton Tableland of northeastern Queensland, and ODP 820, a marine sediment core taken about 45 kilometres offshore from Cairns (Kershaw *et al.* 1993; Moss & Kershaw 2000). There was rainforest at Lynchs Crater until about 78 kyr ago, with rainforest conifers (mainly *Araucaria* and *Podocarpus*) increasing in cooler and drier periods, and flowering plants increasing in warm and wet interglacials (Kershaw 1986; Moss & Kershaw 2000). Beginning at 46 kyr ago rainforests were completely replaced by sclerophyll forest of mixed eucalypt and acacia, in a process that gathered pace between 40 and 35 kyr ago (Turney *et al.* 2001). There was also an increase in Poaceae just before the LGM, indicating an increase in swamp grasses as the lake became more shallow, followed by a decline at the LGM, which could reflect complete drying of the lake. Charcoal had been practically absent from the record until a sharp rise at 46 kyr (Turney *et al.* 2001b).

The ODP 820 core samples pollen that was probably transported in freshwater streams from a large area of the adjacent coast and ranges, including Lynchs Crater. Kershaw *et al.* (1993) provide a general description of changes in pollen in the core over the last 1.5 million years, and Moss & Kershaw (2000) go into more detail for the last 200 kyr. The long record suggests that for most of the last one and a half million years rainforest was the dominant vegetation, with conifers especially abundant. Sclerophyll forests were evidently present and were dominated by casuarinas, with banksias and acacias but few eucalypts until within the last 200 kyr. There were also freshwater aquatic plants, grasses and sedges, ferns, and episodic occurrences of mangroves (presumably indicating high sea levels). Vegetation change, especially the rise of eucalypt-dominated sclerophyll vegetation and decline of rainforest conifers, was concentrated within the last 200 kyr of the sequence. Unlike at Lynchs Crater, however, there were two episodes of change, one in each of the last two glacial cycles. Also, Poaceae continued to increase through the LGM while sclerophyll vegetation declined, indicating the regional development of grasslands or open savannas. Charcoal was continuously present but increased close to the end of each of the last two glacial cycles.

A sediment core from Lake Coomboo on Fraser Island, thought to encompass at least the last four glacial cycles, shows increases in dry forest around the times when the lake was dry, and rainforest or wet sclerophyll when it was full (Longmore & Heijnis 1999). For most of the record this dry forest had a large casuarina component as well as eucalypts and *Angophora*. There is an exception at a drying event dated to about 22 kyr ago, when the lake was surrounded by a eucalypt-dominated dry sclerophyll forest and casuarinas disappeared, to recover strongly during the last 12 kyr. Rainforest was initially well represented, but declined through the whole sequence to be replaced by eucalypt and melaleuca-dominated wet sclerophyll forest. The most pronounced shift from rainforest to wet sclerophyll happened at about 350 kyr. Charcoal was present throughout the sequence, and it increased more-or-less steadily to a maximum value at about 20 kyr.

The Lombok Ridge and Banda Sea marine sediment cores contain pollen from both the Indonesian and Australasian regions. The blending of these pollen sources makes the cores rather difficult to interpret, but Wang *et al.* (1999) argued that the high values for grasses, eucalypts and sedges reflect a predominantly Australian influence. The cores record shifts from ferns, mangroves and rainforest plants in interglacials, to increased sedges and chenopods and grasses during glacials. There was a major and sustained change from about 185 kyr (at the beginning of the coldest and driest phases of the second-last glacial cycle) when grasses increased and eucalypt pollen declined. This shift presumably marks an expansion of open savanna in northwestern Australia, and it coincided with a sustained increase in charcoal. The Banda Sea marine core also shows an increase in grassland and a decline in eucalypts around the LGM (van der Kaars *et al.* 2000).

SOUTHEASTERN AUSTRALIA In the southeast, forests in interglacials gave way to open sclerophyll woodland and treeless herb-fields and grasslands in glacials. In wetter sites such as Lake George (Singh & Geissler 1985) and Darwin Crater (Colhoun & van de Geer 1988) the interglacial forests contained cool-temperate rainforest plants, especially *Nothofagus*, and in drier sites like Lake Wangoom they were dominated by eucalypts with some casuarinas (Harle *et al.* 2002). At Lake George the replacement of woodlands by sparse treeless vegetation in the coldest part of each cycle reached an extreme during the second-last glacial when the area was covered by an almost pure grassland. The vegetation at the LGM was rather more diverse and included a variety of shrubs as well as some grasses. The sclerophyll woodlands were initially dominated by casuarinas but were taken over by eucalypts from the beginning of the last interglacial. Charcoal had been rare in the record until the last cycle, throughout which it was almost continuously present.

THE ARID ZONE Good pollen records are concentrated in the wetter parts of Australia, because continuous pollen deposition is possibly only in wet environments such as the beds of permanent lakes or the sea floor.

Studies of past vegetation in the arid inland suffer badly from discontinuities in the preservation of pollen-bearing sediments. For example, Luly (2001a & b) showed that towards the end of the last glacial cycle the vegetation around Lake Eyre was an open steppe of chenopods, daisies and grasses growing among scattered *Callitris* and casuarinas, but he was unable to find a long and continuous pollen sequence to show how this vegetation had developed.

The GC17 marine core taken from about sixty kilometres off the Pilbara coast of northwest Western Australia (van der Kaars & De Deckker 2002) is therefore very valuable. The Pilbara coast is arid, so GC17 should represent the vegetation over a large area of the dry inland. The core covers the last 100 kyr (with, unfortunately, a gap between 64 and 46 kyr ago). At the beginning of the record the vegetation was much as it is now: open eucalypt woodland, but with plants such as ferns, palms and aquatic species indicating a comparatively high rainfall. Moisture-loving plants declined from 100 to 40 kyr ago and trees evidently became sparser. There was a major change soon after 45 kyr ago when chenopod shrubs increased. Chenopods declined somewhat and grasses increased close to the LGM. Rates of pollen input, which broadly indicate vegetation density, were highest between 75 and 80 kyr ago, declined to a low point at about 20 kyr ago, and by 5 kyr ago had recovered to the levels of 80 kyr ago. Charcoal concentrations were relatively high in the first part of the record and low in the second. In the arid northwest, studies of phytoliths – distinctive silica particles found in some plant tissues, which are well preserved in dry sediments – from the archaeological dig at the Carpenters Gap Rock-shelter show an increase in spinifex and a decline of other grasses about 33 kyr ago (Wallis 2001, Plate 10).

New Lands – the Carpentarian Plain As sea levels fell during the long glacial periods of the Pleistocene more dry land was exposed around the coast. The Carpentarian and Bassian Plains re-emerged and there were major extensions of dry land in northeastern Queensland and around the Great Australian Bight. What were these new lands like? The vegetation of the Carpentarian Plain has been inferred from pollen in sediment cores from what is now the sea floor in the Gulf of Carpentaria (Chivas *et al.* 2001; Torgersen *et al.* 1988). It was mainly grassland, with swamp plants growing in low-lying areas and a very open woodland of *Callitris*, *Casuarina* and Myrtaceae (presumably including eucalypts) on the higher and better-drained areas. Streams from southern New Guinea would have flowed onto the plain, and at its centre was Lake Carpentaria, a vast shallow fresh-to-brackish lake with a surface area of more than 30

000 square kilometres (Yokoyama *et al.* 2001). The vegetation of the western side of the Carpentarian Plain can be interpreted from the Lombok Ridge marine sediment core as mainly a grassy eucalypt woodland with large stands of mangrove along the coast. The topography of the sea floor in this area includes several small enclosed catchments which may also have formed freshwater lakes (Yokoyama *et al.* 2001).

More evidence on the environment of the Carpentarian Plain comes from Aru Island, 150 kilometres south of New Guinea, which at the LGM was part of a dissected limestone plateau on the western edge of the plain. Lembudu Cave on Aru has a human occupation sequence dating from about 27 to 12 kyr ago (O'Connor *et al.* 2002). The people who lived there presumably came either from southern New Guinea or northern Australia (the archaeological evidence points to northern Australia; Golson 2001), and their presence confirms that large areas of the Carpentarian Plain were fit for people. The mammals they hunted were agile wallabies, red-legged pademelons, northern brown bandicoots and long-nosed echymiperas *Echymipera kaluba* (a spiny bandicoot, now restricted to New Guinea). These are all species associated with high-rainfall savannas and rainforest edges.

AND THE BASSIAN PLAIN We have good descriptions of the LGM vegetation at several sites in southern Victoria and western Tasmania and on King Island, which indicate the nature of the vegetation over much of the Bassian Plain. At Lake Wangoom and E55-6 there was a change from eucalypt forest with some cool-temperate rainforest elements and little grass at 100 kyr, to an open steppe with a high diversity of shrubs, grasses and non-grass herbs under just a few scattered eucalypt and casuarina trees by 20 kyr ago. The E55-6 record also has some pollen of aquatic plants at the LGM, indicating swamps and wetlands (Harle 1997). The Wyelangta record samples what is now an isolated patch of cool-temperate rainforest of Antarctic beech *Nothofagus cunninghamii* in a high-rainfall location (Mckenzie & Kershaw 2000). Here, *Nothofagus* was continuously present through the LGM, although low pollen concentrations suggest that tree cover became sparse. The presence of sedges and heath plants along with grasses indicate that the site became swampy during the LGM. Pollen records from western Tasmania show that at that time an open steppe composed almost entirely of grasses and daisies, with just a few alpine shrubs, had replaced the *Nothofagus* forest that had dominated previously; eucalypt and *Nothofagus* forests persisted only as isolated remnants on valley floors close to the coast (Colhoun 2000). At Egg Lagoon on King Island closed forest at 120 kyr was gradually

PLATE 13 Spiny foliage of juvenile plants in two arid zone acacias: **(a)** a juvenile *Acacia peuce* showing its typical defensively spiky hummock form, **(b)** pungent tips of the stiff individual phyllodes in *A. peuce*, and **(c)** juvenile foliage of *A. pickardii*. (photographs by Jon Luly)

PLATE 14 Height dimorphism in *Acacia peuce*, in which foliage within reach of large terrestrial herbivores is defended and foliage out of reach is undefended: **(a)** a half-grown plant, with dense spiny foliage (as in Plate 13) low down, and open soft foliage above; **(b)** mature form – a tall erect tree with undefended foliage. (photographs by Jon Luly)

PLATE 15 Dry rainforest, a biodiverse and formerly widespread plant community of northern Australia in which many plants show signs of having co-evolved with Pleistocene megafauna.

PLATE 16 A possible 'megafauna fruit', *Capparis canescens*, the wild orange (or wild pomegranate) of central and northern Queensland. The fruit is about the size of a small orange, with a tough rind and small seeds distributed through a sticky pulp. (photograph from Anderson 1993, reproduced courtesy of Eric Anderson)

replaced by sclerophyll forest with an increasing Casuarinaceae element, until about 35 kyr ago, when forests were replaced by herb-fields and grasslands (D'Costa 1997).

So, the most widespread vegetation type across the now-submerged Bassian Plain was probably a treeless mixed shrubland and grassland. Isolated forest stands held on in some humid sites, and there are signs in the pollen record of wetlands. A basin in the centre of the region, enclosed by eastern and western highlands now represented by Flinders and King islands, was probably a large freshwater lake, and there are many scattered depressions elsewhere that may have been lakes or marshes (Lambeck & Chappell 2001).

CARBON ISOTOPE SIGNATURES OF VEGETATION CHANGE

Most shrubs and non-grass herbs, and all trees, use the C3 photosynthetic pathway, and as a result their tissues are more depleted in ^{13}C than plants with C4 photosynthesis (mainly grasses, especially warm-climate species). Changes in the concentration of ^{13}C in fossil material can therefore be used to detect broad trends in vegetation structure. The ^{13}C concentrations in soil organic matter in the Gregory Lakes system in the arid north show a steady increase from about 120 kyr until the recent, indicating a change from a near-pure C3 vegetation to one composed of about 75 per cent C4 plants (Pack *et al.* 2003). This could have been a response to increased aridity through the last glacial cycle. Van der Kaars *et al.* (2000) analysed the isotope composition of charcoal in the Banda Sea marine core to reconstruct changes in the vegetation that was being burned in the source regions for this core, and found a quite sudden shift from predominantly C3 to predominantly C4 at around 35 kyr. The abruptness of the change might be at least partly a function of the location of the Banda Sea core on the sea bed to the north of Australia. The fall in sea level at about 35 kyr would have increased the area of northern Australia within the source region of the core and thereby raised the contribution of dry (and predominantly C4) Australian grasslands.

Both records, especially the Gregory Lakes one, are interpretable as representing a climate-caused expansion of grasslands, but this is complicated by the fact that many chenopod shrubs, especially the more arid-adapted species, also have C4 photosynthesis (Sage *et al.* 1999) and could have contributed to the isotopic change. The GC17 core shows chenopods increasing from about 45 kyr, and grasses increasing much closer to the LGM. A possible interpretation of this is that the increase in C4 plants was initially due to the expansion of chenopods, and then later to C4 grasses that were favoured by the aridity and low CO_2 of the LGM climate.

Another important data-set on changing isotope composition of ecosystems has been provided by Miller *et al.* (2005), who measured the carbon isotope composition of emu and *Genyornis* eggshell and used it to infer the vegetation available to those birds through the last 140 kyr. This is discussed in Chapter 7.

SYNTHESIS

The pollen data summarised above should be interpreted cautiously, for two reasons. First, the dating of many of these cores is uncertain. All can be dated by radiocarbon in their upper levels, but this is possible only for the last 45–50 kyr at most. Beyond this, dates are usually inferred by reference to standard marine oxygen isotope chronologies. Marine cores can be directly calibrated against these standard chronologies using oxygen isotope measurements on microfossils in the cores themselves. The dating of terrestrial cores is much more tenuous, and at deeper levels usually depends on the matching of vegetation changes reconstructed from the pollen in the cores to climate changes represented in oxygen isotope chronologies. A certain amount of guesswork is involved in this matching – not to mention circularity, when the goal is to infer the effects of a changing climate on vegetation. The approach does not provide good chronological control for events that were not closely aligned to distinct changes in the isotope record, and discontinuities in the sediment records may be missed. In some cases (Lynchs Crater, Lake Wangoom) it has been possible to buttress chronologies of terrestrial cores by matching them to pollen profiles in nearby and more securely dated marine cores. Absolute dating by techniques such as U/Th has helped to fix dates in some terrestrial records but has not been applied at the intensity needed to produce precise chronologies independent of the oxygen isotope standard.

A second problem is the lack of coverage of the dry inland. We know very little about the vegetation history of the dry tropics of inland northern Australia, the southern arid inland, or southwestern Australia. The carbon isotope studies go some way to filling gaps in the pollen data, but they would be more powerful if they could be matched to pollen records showing which plant taxa were responsible for producing the observed changes in isotope ratios. The dating of the Gregory Lakes isotope sequence is too coarse to resolve the tempo of change through the last glacial cycle, and the Banda Sea record is confounded by shifts in the dominant geographical source of pollen.

Despite these problems, some broad patterns in vegetation change through the last glacial cycle are clear. During the last interglacial, about 130 kyr ago, forest cover was more widespread than in recent times. In particular there was a greater extent of rainforest, including distinctly more conifer-dominated dry rainforest. Wet forests retreated with the steep fall into glacial climate conditions between about 130 and 110 kyr. The pace of vegetation change then slowed from 110 kyr until about 35 kyr ago, when the climate stepped down again to substantially cooler and drier conditions. At that time much of the remaining closed forest was replaced by dry open woodlands; elsewhere sclerophyll forests and woodlands gave way to grassland or mixed grass/shrub steppes. Cold alpine grass/shrub steppe covered the ranges of southeastern Australia. Tropical wet rainforest, which disappears completely from the fossil record for a few thousand years, must have contracted to very small remnant fragments, while wet sclerophyll forests and cool-temperate rainforests persisted in a precious few refuges near the coast. Rainforest conifers tended to become rare during the last cycle compared with the preceding ones, and in some places casuarinas were replaced by eucalypts as the dominant trees in dry sclerophyll forest.

The weakening of the tropical monsoon meant that landscapes in the north became uniformly arid, and tropical grasslands, woodlands and dry rainforest thickets were largely replaced by a treeless mixed shrub/grass steppe. Northern Australia as a whole must have looked like much of the arid inland of southern Australia today, particularly in the wide occurrence of low shrublands and the scarcity of water. On the other hand, southeastern Australian landscapes were 'very different from any modern analogue: active dunes and dust entrainment from a sparsely vegetated land surface, but with large rivers and enhanced runoff from the highlands' (Hesse *et al.* 2004), as well as stable wetlands in some areas that are now dry.

Plant growth would have been strongly constrained by the low temperature, rainfall and atmospheric CO_2 of the LGM, and the dynamics of vegetation would have slowed down. Successional changes in vegetation, growth flushes in good seasons, and recovery from disturbances such as fire would all have been subdued. Australian landscapes must have been very uniform and unchanging under these conditions, with fewer contrasts in space and time than we are familiar with now. The Australian landscape is often described as 'timeless'. No one who is aware of the dramatic changes in Australian environments over the last one hundred thousand years could possibly say this, but for the twenty thousand years of the LGM the continent must have had a distinctly timeless quality.

PEOPLE, FIRE AND VEGETATION

There is no doubt that fire was a significant environmental factor in Australia before people came. There is charcoal in sediments millions of years old, and plant groups like *Banksia* and the eucalypts have many adaptations to fire (Bowman 2000; Kershaw *et al.* 2002). But there is a quite general view that with the arrival of people, burning became much more widespread and frequent, and caused the replacement of fire-sensitive by fire-tolerant vegetation over much of the continent. This was initially inferred from evidence of the pervasive effects of Aboriginal burning on the distribution and structure of vegetation at the time Europeans arrived. Projecting these effects back in time, it seemed inevitable that the first arrival of people caused major changes to vegetation. A dramatic impact of human arrival seemed to be confirmed when Singh *et al.* (1981) described two late Pleistocene pollen cores, from Lake George and Lynchs Crater, with abrupt increases in charcoal and dramatic vegetation changes that they argued were due to the arrival of people.

There has been a great deal of work on the Pleistocene history of fire in Australia since Singh *et al.* (1981) published their hypothesis. This work has been reviewed by Kershaw *et al.* (2002), who summarised developments in the palynological (fossil pollen) evidence, and by Bowman (1998), who provided an essential overview of the ecological, anthropological and palaeoecological aspects of the problem. The most general conclusion from this work is that there was *not* an unprecedented and continent-wide increase in burning, with associated vegetation change, when people first came to Australia. This is illustrated in Figure 5.4a, which shows major changes in charcoal in the long pollen cores summarised above. All the cores show pronounced shifts in charcoal concentrations through time, indicating changes in fire regimes. These changes usually coincided with vegetation transitions, such as from a rainforest to sclerophyll forest environment at Lynchs Crater, and from sclerophyll woodland to open grassy savanna in the Lombok Ridge record. But, of the five records that cover most of the last two glacial cycles, all but Lynchs Crater had episodes of high charcoal in both (although the episode in the second-last cycle at Lake George was brief). Within glacial cycles, increases in charcoal tended to coincide with periods of declining temperature (and presumably rainfall). Nonetheless, dates for charcoal increase during the last glacial period vary widely from one core to another. At E55-6 and GC17 charcoal was highest early in the cycle, at Lake Selina and Lynchs Crater it increased in the second half of the cycle,

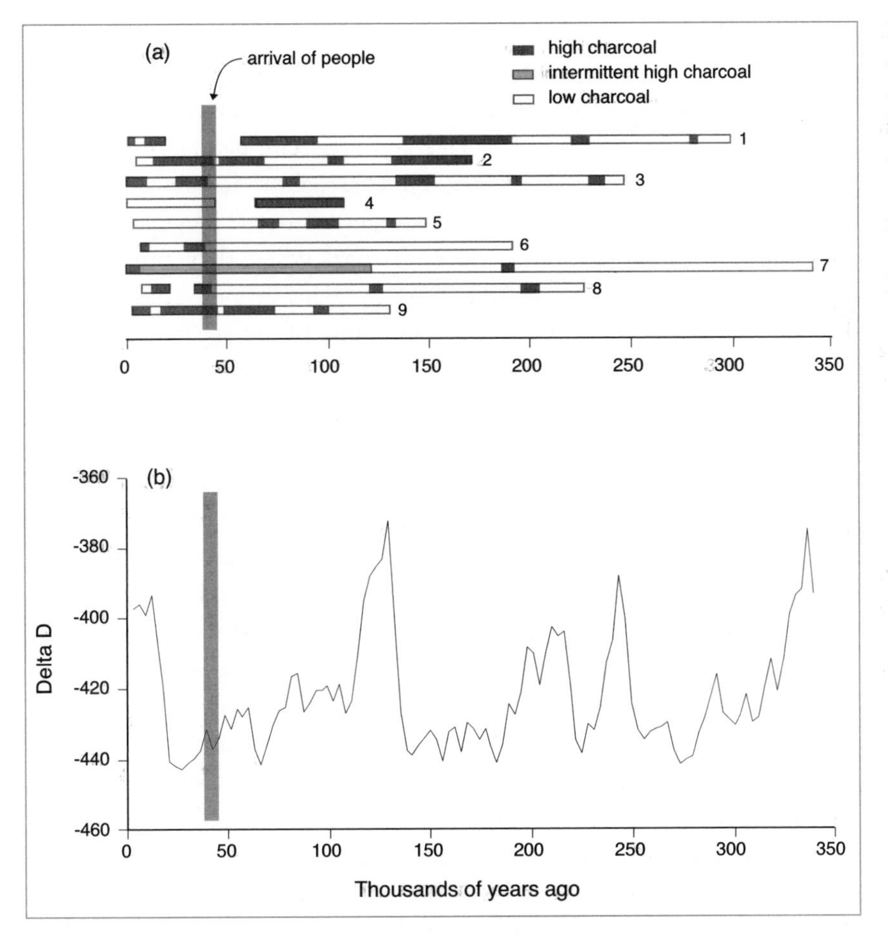

FIGURE 5.4 Summary of changes through time in charcoal concentrations from long-term sediment cores (for locations of cores, see Figure 1.2) in relation to climate change: (a) episodes of high and low charcoal through time in nine well-dated sediment cores; (b) changes in atmospheric temperature through the last three glacial cycles, as indicated by deuterium concentrations in Antarctic ice (EPICA 2004).
Notes: 'High' charcoal is charcoal above the mid-range of values for the whole of each record. The records are Lombok Ridge (1), Banda Sea (2), ODP 820 (3), GC17 (4), E55-6 (5), Lynchs Crater (6), Lake George (7), Lake Wangoom (8), Lake Selina (9). Cores 1–5 are marine and 6–9 are terrestrial. High points on the graph indicate high temperatures, which were associated with high rainfall. The vertical bar indicates the period during which people are most likely to have arrived and become widespread.

and at Lake George it was intermittently high throughout the last glacial cycle.

Four other long-term records were not included in Figure 5.4 because of uncertainties over their dating, but they reveal still more variation in charcoal history. At Darwin Crater in southwestern Tasmania charcoal was high during glacials and low during interglacials, with no underlying trend through the whole record (Colhoun & van de Geer 1988). At Old Lake Coomboo on Fraser Island there seems to have been a steady increase in charcoal over perhaps the last 600 kyr to a peak at about 20 kyr ago. At Egg Lagoon on King Island charcoal values were low from the beginning of the record at about 130 kyr ago before increasing at about 60 kyr ago, or perhaps a little earlier (dating of this record between about 74 and 35 kyr ago is particularly uncertain, D'Costa 1997). Analysis of pollen at Cuddie Springs identified a charcoal peak at about 28 kyr that was not associated with a major vegetation change. The record shows an earlier charcoal peak that coincided with a decline in *Casuarina* and an increase in chenopod shrubs; when this happened is unknown, but it was evidently well before the arrival of people (Field *et al.* 2002).

These data on charcoal suffer from the same problems identified above for pollen: poor geographical coverage and uncertain dating. Another difficulty of interpretation is that changes in charcoal can be viewed either as a cause or a consequence of shifts in vegetation state. Vegetation types with higher biomass produce more charcoal when they burn, so the replacement of a light vegetation type by a heavy one could lead to increased charcoal even if fire was not the cause of the vegetation change. This can explain why the transition from cool-temperate grassland to eucalypt forest during glacial periods in southern Australia was accompanied by increased charcoal, as at Lake Wangoom. The fact that terrestrial records are mostly from moist microenvironments, especially the beds of perennial lakes and swamps, makes interpretation of fire history even more difficult, because such sites are likely to have been buffered against effects of increased burning in the surrounding landscape and so might record a delayed signal of those increases, if they registered them at all.

The marine records are particularly valuable because they reflect broader patterns of burning over large source areas of adjacent mainland regions; also they are more securely dated. They have an unavoidable bias towards coastal regions, but they do sample a range of climate zones: the arid northwest (GC17), the wet-dry tropical north (Lombok Ridge and Banda Sea), the wet tropical northeast (ODP 820) and the dry-temperate south (E55-6). There are common features in these records that are not reproduced so clearly in the more localised terrestrial records. All those that reach back to the second-last glacial cycle show high charcoal levels near its end, and all show episodes of high charcoal in the early part of the last cycle, between 100 and 60 kyr ago.

The strongest evidence for an impact of human arrival comes from Lynchs Crater, where charcoal levels were very low until an increase beginning at about 46 kyr ago. This date also marks the beginning of a decline in rainforest, especially of rainforest conifers (Kershaw 1986; Turney *et al.* 2001b). Lynchs Crater is the best-dated and most thoroughly studied record of late Quaternary vegetation in Australia, and it is widely interpreted as showing an impact on vegetation of increased burning by Aboriginal people. But the major vegetation change at this site was concentrated between 40 and 35 kyr ago, five thousand years or more after the arrival of people, and no other site shows such an unprecedented change. Even ODP 820, which samples a larger section of the same part of northeastern Australia, reveals an increase in charcoal and decline of rainforest in each of the last two glacial cycles. The only other record with a rise in charcoal at or soon after 45 kyr is Lake Wangoom. But this charcoal peak marks a very strange moment in that record: very little pollen was being deposited, and what little there was came from plants that had been extinct for millions of years. Harle *et al.* (2002) suggest that this pollen came from the Tertiary sediments of the crater wall that were being redeposited in the bed of the lake, possibly as a result of volcanic activity, which might also explain the rise in charcoal relative to pollen. The lack of association between human arrival and fire increase is also shown in Figure 5.5, which summarises the timing of peaks in charcoal identified by Kershaw *et al.* (2002) in seven long-term records that they considered to have secure chronologies. The period between 40 and 70 kyr ago is revealed as one with unusually few abrupt changes in burning.

Much of the variation in charcoal concentrations summarised in Figure 5.4 can be explained ultimately as a function of climate, without invoking people. During glacial periods vegetation became more flammable as a result of the drier climate. Burning would have increased as a simple result of a drier vegetation interacting with natural sources of ignitions (presumably lightning), but fire probably contributed to further vegetation change because the plants that regenerate well after fire are often themselves highly flammable, and so promote further fires and still more change towards a fire-prone landscape. The link between charcoal and climate might emerge more strongly if variation through time in the strength of the ENSO system could be taken into account. Kershaw *et al.* (2003) showed that charcoal peaks in the ODP 820 record are closely related to peaks in ENSO activity predicted from a global climate model (Clement *et al.* 1999). There is a compelling suggestion here that charcoal was especially high when a strong ENSO coincided with late-glacial arid conditions, but to be sure of such a link we would need to relate charcoal

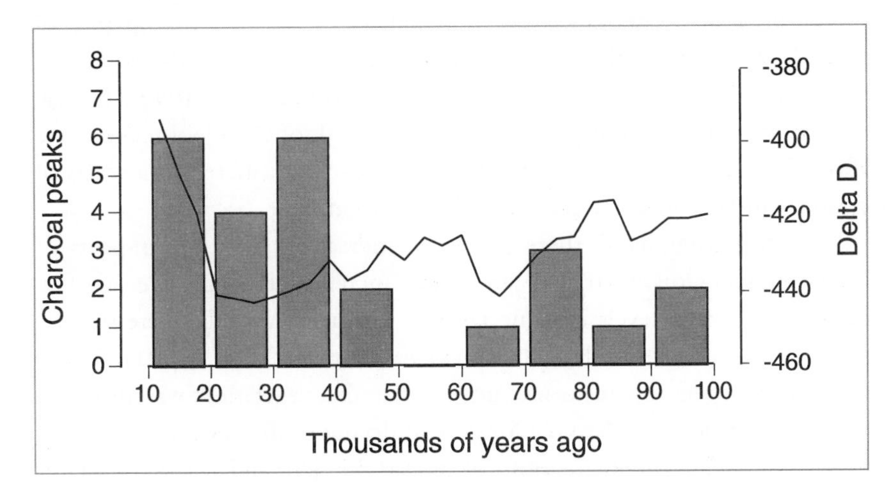

FIGURE 5.5 Number of charcoal peaks through time (bars) in well-dated sediment cores, from Kershaw *et al.* (2002), in relation to temperature change (continuous line) over the last 100 kyr. *Notes:* Changes in atmospheric temperature are indicated by deuterium concentrations in Antarctic ice (EPICA 2004), where high points on the graph indicate high temperatures.

levels to empirically derived estimates of the strength of ENSO, and there is too little data for this at present.

The strongest signal in the charcoal data is that there was more charcoal in the last glacial cycle than in the preceding one, and that charcoal tended to increase through the last glacial cycle (Figure 5.4). Burning appears to have been much more widespread through most of the last 10 kyr (the present interglacial) than in the previous interglacial about 130 kyr ago. There seems no doubt that fire has increased in Australia over the last 120 kyr, but the arrival of people in the middle of the last glacial cycle had very little impact on this trend (although I will argue in Chapter 8 that human influence on burning patterns strengthened within the last few thousand years).

THE LAST GLACIAL CYCLE IN PERSPECTIVE

There is no doubt that the last glacial cycle, and in particular the LGM, caused dramatic changes to Australian landscapes. But Australia had already passed through many glacial cycles. Was the last one the worst?

The record of global temperature change from Antarctic ice cores does not show that the last cycle was unusual, nor does it reveal any trend in mean temperature or in the amplitude of temperature cycles over the last half million years. The second-last glacial cycle looks more extreme than the last one in that it had a longer interval of very low temperature (Figure 5.1), which coincided with an extended period of dust accumulation in marine sediments (Figure 5.2). The dust record certainly does not show that vegetation cover was more sparse at the LGM than in the preceding glacial maximum. Reconstruction of the past topography of Lake Eyre suggests that it was more arid at the second-last glacial maximum 140 kyr ago than at the LGM (Miller *et al.* 1999). It may actually be that the most profound climate shock of the Pleistocene came about 700 kry ago, when deep temperature cycling ushered in a series of episodes of very low temperature and deep aridity.

On the other hand, there are some indications that the impacts of glacial climates on Australian landscapes increased through the last few cycles. Peak water levels in some lakes in northern Australia were successively lower over the last three cycles (Bowler *et al.* 2001). Three long-term vegetation histories suggest that the vegetation was driest in the last cycle: at Fraser Island a long-term drying of the vegetation appears to have underlain the fluctuations of the last 350 kyr; and at Lynchs Crater and ODP 82 there was less rainforest and more sclerophyll forest at the last than in the preceding glacial maximum. Other vegetation histories, particularly Lake George and Darwin Crater, imply that the vegetation was more arid in the last two glacial cycles than before, without showing that the last cycle was drier or colder than the second last. Further, although it is very difficult to sustain the case that the arrival of people was followed by an increase in burning, long-term charcoal records do suggest that there was more burning through the last glacial cycle than in the one before.

So, the evidence on the relative severity of the last glacial cycle is somewhat confused. It is difficult to make a case that the climate of the LGM was more extreme than at preceding glacial maxima, but it can still be argued that the impacts of aridity on landscapes became successively greater with each of the last few glacial cycles. It is possible that the extra climatic ingredient forcing landscape change through the last few glacial cycles was climate variability due to ENSO. ENSO cycling of rainfall creates the conditions for increased fire, because vegetation builds up during wet phases and then dries out and provides fuel for fires during the long droughts that follow. Soil erosion also increases when heavy rains fall on landscapes that have passed through years of drought and fire.

Further, vegetation states of low and high flammability may be separated by distinct thresholds that are more easily crossed in the direction of increased burning than the reverse. For example, *Allosyncarpia ternata* forest in northern Australia suppresses fire because the tree's horizontally arranged leaves intercept so much light that a continuous grass layer is unable to develop on the forest floor. An opening up of the canopy, which could initially be due to a drier climate, allows the development of a grass layer that can carry fires, and this excludes *A. ternata* because it cannot regenerate in the presence of fire. Eucalypts can regenerate with fire, and their pendulous leaves form an open canopy under which the grass layer persists and fires recur, holding the vegetation in its new state (Bowman 2000). Through mechanisms like this, landscapes may be locked into a state of high fire frequency on fire-tolerant vegetation. These transitions may have been made at different times in different places, but their cumulative effect would have been to create a generally more flammable and drier environment.

6

Testing hypotheses on megafauna extinction

THIS CHAPTER BRINGS together the evidence that can help us decide on the causes of the megafauna extinctions. For philosophical reasons given at the end of Chapter 3, I treat each of the three major hypotheses separately. The arrangement of evidence follows the plan that was laid out in that chapter. That is, each hypothesis will be tested according to (1) how well it fits the timing of extinctions, (2) whether it can explain why it was that certain species went extinct while others survived, and (3) whether it could have been powerful enough to account for the magnitude of the extinction event.

TIMING OF EXTINCTIONS

Our current understanding of the timing of the megafauna extinctions is that they were concentrated between 50 and 45 kyr ago, or at least that there was an abrupt decline in geographic range and abundance of large vertebrates at that time (see Chapter 4). If climate change wiped out the megafauna, the process should have begun with Australia's fall into the deep aridity of the LGM around 35 or 30 kyr ago, and it should have been completed by about 20 kyr ago. Before 40 kyr ago there were several alternations between moderately arid and moderately humid conditions, but none of these was exceptional; in fact they were typical of the climate of most of the preceding million years. A disappearance of megafauna around 45 to 50 kyr ago is too early to have a climatic cause, particularly as this was a time when the climate had ameliorated somewhat after an earlier arid phase (Figure 5.1).

Climate-change models predict that the megafauna should have disappeared first from the centre of the continent and survived longest near the coast, because deep aridity took hold first in the centre and then expanded towards the continental margins. This prediction can be tested with the full set of dates on megafauna remains described in Table 4.2. Johnson (2005b) used these data to plot the most recent dates for megafauna remains at late Pleistocene sites in relation to the distances of those sites from the present-day coast. Whether this analysis is valid depends on the reliability of the evidence for survival of megafauna past 46 kyr and, as discussed in Chapter 4, there are serious doubts over most of that evidence. Nonetheless there is no hint in these dates of a wave of extinctions travelling from the centre to the edge of the continent.

The timing of increases in burning and associated vegetation change in the last glacial cycle varied a great deal across Australia, and there is no direct evidence for a coincidence of extinction with increased fire. In southern Victoria, for example, it seems that changes in fire happened early in the last glacial cycle (see Figure 5.4), to the extent that they happened at all, but many species of megafauna survived this period only to go extinct in the latter part of the last cycle.

On the other hand, the association of megafauna extinction (or dramatic decline) at 50–45 kyr ago with the conservative date of 45 kyr ago for human occupation of Greater Australia is very close. The fit gets closer still if, considering the incompleteness of the fossil and archaeological records, we allow that the disappearance of the megafauna happened a bit later, and the arrival of people a bit earlier, than has been revealed by dating studies so far.

A possible exception is Tasmania, where Cosgrove & Allen (2001) suggested that megafauna may have gone extinct before the arrival of people, because late Pleistocene occupation sites in southwestern Tasmania contain no megafauna species among very large collections of identifiable mammal bone. However, this difficulty may be more apparent than real, for the following reason. If people did hunt the megafauna to extinction, most of the damage would have been done within the first thousand years after human arrival (see below). Bone is comparatively rare in this early interval. Of five Pleistocene cave sites reviewed by Cosgrove & Allen (2001) only one contains any bone dating from the first 4 kyr of human presence in Tasmania, and this early sample represents only about 0.006 per cent of the total bone fragments. The power of these data to detect human–megafauna overlap may be very low if megafauna species went extinct within a few thousand years of human arrival. As an illustration of this problem, we know that Tasmanian Aborigines hunted grey kangaroos, but remains of this species are extremely rare in Cosgrove and Allen's sample of bone, with none from the first 4 kyr of human occupation. Had Tasmania's eastern grey kangaroos gone extinct then, we would be left with no evidence that they overlapped with people.

THE LIVING AND THE DEAD

The evidence contained in the selectivity of extinctions has been largely ignored in the extinction debate. Mostly the selectivity of extinctions has been invoked to buttress hypotheses rather than to test them. Typically someone will identify a single trait thought to have been shared by the extinct species (such as that, as large animals, they needed to drink a lot), build an extinction scenario around that trait, and then re-emphasise the trait's significance to strengthen the original hypothesis. This style of argument ignores the many other features of species that might affect their ability to survive a given threat and that could be used in testing hypotheses, and it makes almost no use of the species that survived.

A more productive analysis would go in the opposite direction: we should start with a hypothesis on what might have caused extinctions in the late Pleistocene, think of the traits that would make species more or less likely to survive that factor, and then test if the extinct species and the survivors differed in those traits in ways predicted by the hypothesis.

CLIMATE CHANGE

Given what we know of the character of climate and vegetation change in the late Pleistocene, what traits should have made species vulnerable to it? Here is a short list of the most obvious ones:

Dependence on dense forest There was a dramatic reduction of tree cover at the LGM, and wetter forest types survived only as small isolated remnants. Where forest persisted, the trees were evidently more widely spaced and probably smaller than now. Wet-forest mammals, especially those that lived in the forest canopy, must have suffered very large population reductions as a result. Gliding possums should have been most sensitive to these changes because they avoid coming to the ground and rely on close spacing of trees to move through their habitat.

Dependence on dense and diverse ground vegetation The coldness, aridity and CO_2-scarcity of late-glacial climates produced open and sparse vegetation with low ground cover. Species that relied on dense vegetation at ground level to hide from predators and to avoid exposure to extreme environmental conditions – that is, most small mammals – should have suffered as a result. Various species of small mammals often have distinct requirements for vegetation structure, so that a diversity of vegetation types is needed to maintain high species diversity. The sparseness and the uniformity of vegetation at the LGM should have produced simpler communities of small mammals, with the loss of many habitat specialists.

Requirement for high-quality plant food Under the conditions of limited plant growth in the last glacial, the nutritional quality as well as the density of vegetation would have declined. This would have been bad for herbivores that needed to feed selectively on foliage with high concentrations of nutrients. There is a strong relationship between body size and the nutrient quality of plant food needed by herbivores. Metabolic requirements per unit body mass decline with increasing body size, which means that small-bodied herbivores need to select food of higher nutritional quality. Species like the hare-wallabies and pademelons feed selectively on nutrient-rich plant parts like growing shoots, while larger kangaroos feed in bulk on mature leaves and stems. The smaller species would therefore have been most strongly affected by a fall in the nutritional quality of plants.

Insectivory The very smallest terrestrial mammals are insectivorous, and their food intake is strongly limited by temperature because insect activity declines at low temperatures. Small dasyurids – planigales, dunnarts, ningauis and others – cope with fluctuating prey activity by entering torpor when temperature and food availability are low (Geiser 2004). This is an effective strategy for riding out daily fluctuations in insect activity, but these species do not hibernate long-term and so have no way to avoid the consequences of long periods of low temperature and non-availability of their insect prey. The long periods of unrelieved cold in the LGM would have made life very hard for them.

It is striking that there were very few extinctions of mammals with these characteristics. About 28 per cent of living Australian and New Guinean marsupials are arboreal, but only one (1.8 per cent) of the extinctions was of an arboreal species. This was a large koala (perhaps a large tree kangaroo *Bohra paulae* as well, but there is very little fossil material for this species and what there is comes from the early Pleistocene). Evidently, no gliders were lost. The survival of rainforest-dwelling arboreal species is remarkable. Only one mammal with a mass of less than one kilogram (*Antechinus puteus*) is known to have gone extinct; this was also the only recorded extinction of a small insectivore, and small-bodied herbivores survived well. While it is possible that extinctions in small mammals and arboreal species have been underestimated because their remains are less well preserved as fossils, it is unlikely that new discoveries will reverse the conclusion that extinction rates were relatively low in these groups.

The species that should have been most resilient to the stresses imposed by glacial climates were large-bodied herbivores. There are several reasons for this. Because mass-specific energy requirements decline with body size, large herbivores can feed unselectively on low-quality foliage. Large size also provides them with a capacious digestive tract, both in absolute volume and in volume relative to total energy requirements. Parts of the digestive tract serve as fermentation chambers in which structural carbohydrates in plant tissue are converted to useable energy by symbiotic microbes. This allows the herbivore to feed in bulk on low quality food and to cope with declining quality of foliage by increasing its total food intake. The largest living kangaroos and wombats have very specialised digestive systems of this kind (Hume 1999) and no doubt the giant extinct species were even better equipped in this way.

In addition, the range of movement of individual herbivores increases with body size, and does so at a rate *greater* than their whole-body energy requirements (Figure 6.1). This gives large herbivores an advantage over small ones under conditions of increasing sparseness of food availability, because they are able to move through larger areas to find the energy they need. The same applies to water. Large mammals' increased water requirements are more than compensated by their capacity for long distance movement. Water sources might have become more widely scattered under glacial climates but the ability of mammals to cope with this by moving between water points increases with body size (and anyway there is evidence that in some places the availability of drinking water actually improved – see chapter 5).

It could be argued that large species would have been most sensitive to reduction in the area of their habitat, because they typically have low population densities and therefore need large areas to maintain viable populations. This idea could work for forest-dependent species such as tree kangaroos because there is no doubt that their habitats did contract, but there are problems applying it to most of the extinct megafauna. The argument assumes that the environmental changes of the last glacial would have severely restricted the distribution of large-bodied species – but, as pointed out above, large species should have been especially resilient to them.

In any case, for most of the last glacial cycle, open shrublands, grasslands and woodlands were expanding as a result of climate change (see Chapter 5), and this should have increased the distributions of very large herbivores. It was only in the depths of the LGM that these habitats might have been replaced by dry deserts, but it seems that the megafauna had already disappeared before this happened. We also need to recall that as aridity took hold in the interior of the continent, Australia was getting bigger because of drastic falls in sea level. The new habitat being created at the margins

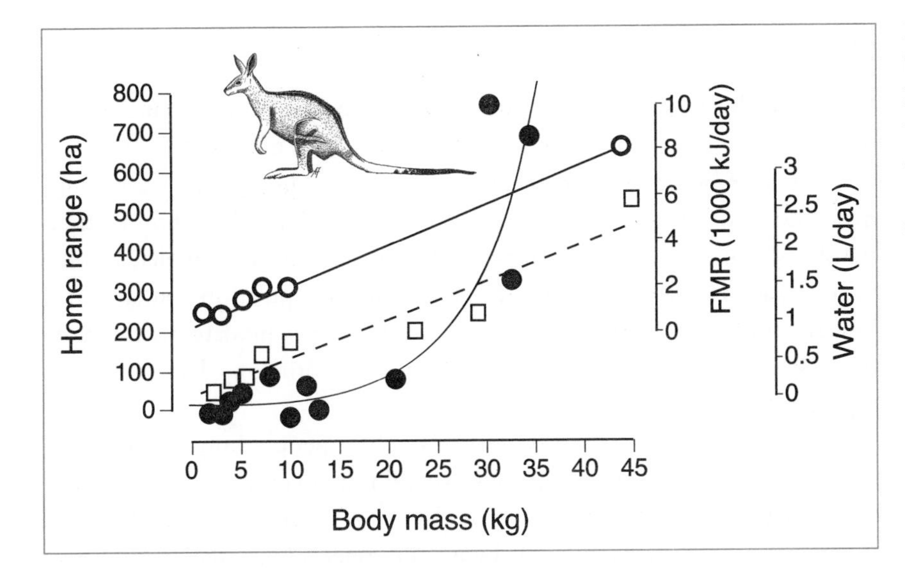

FIGURE 6.1. Relationship between body mass and home range size (●), water requirement (❑) and field metabolic rate (○) in macropodid marsupials.
Notes: Data on home range from Johnson (1998), and on water and field metabolic rate from Hume (1999).

consisted of more open shrublands, grassland and woodland growing over vast and well-watered plains that should have suited species such as the *Sthenurus* kangaroos and *Diprotodon optatum*. Probably, total habitat area for many of these species was actually increasing as they went extinct.

The one feature of the selectivity of late Pleistocene extinctions that has been used to support environmental change as a cause is that so many of the extinct large marsupials were browsers. Presumably they relied on a dense layer of shrubs and small trees in their habitat, and that layer, conceivably, disappeared as a result of climate change or fire (see Chapter 3).

The problem with this argument was pointed out by Prideaux (2004) and Johnson & Prideaux (2004). It is true that most of the extinct species were browsers (see Table 2.1), but this is a statistical consequence of the fact that browsers predominated among the very largest species, combined with the relationship of extinction rates to body size. The large species that were grazers also went extinct, and the relationship between body size and extinction was quantitatively similar for browsers and grazers. The similarity of the extinction pattern for grazers and browsers once body size has been taken into account actually provides further evidence that environmental change was not part of the cause of their extinction. If, for example, there was an increase in fire and it destroyed the complex mixed woodland/shrubland habitats that might have been needed by large browsers, these habitats would have been replaced by grasslands, which should have been good for large grazers. In general, if an environmental shift was the cause of the extinctions, one would not expect species with contrasting ecological requirements to have declined in such similar ways.

OVERKILL

If hunting by people caused extinctions in the late Pleistocene, species with the following characteristics should have been most at risk:

Low reproductive rate Slow-reproducing species are typically also long-lived and have long generation times. This collection of life-history traits translates into a low maximum rate of population growth, which in turn means that the maximum rate of harvest that a population can sustain is low. An increased mortality rate imposed by a new predator is therefore more likely to cause the extinction of species with low than with high reproductive rates.

Living in open habitats, on the ground Because people hunt primarily by sight and from the ground, ground-dwelling species of open habitats would have been most exposed to predation by humans. Species that lived in dense and visually occluded habitats, and especially in trees, would have enjoyed some protection.

These two predictions are confirmed in Figure 6.2, which plots extinction rates in families of marsupials against the average reproductive rate in each family (reproductive rates for extinct species were estimated from relationships between body mass and reproductive rates among their living relatives). There were few or no extinctions in groups in which average reproductive rates were substantially greater than one offspring per female per year, but at lower reproductive rates extinction increased to 100 per cent. There were arboreal as well as terrestrial

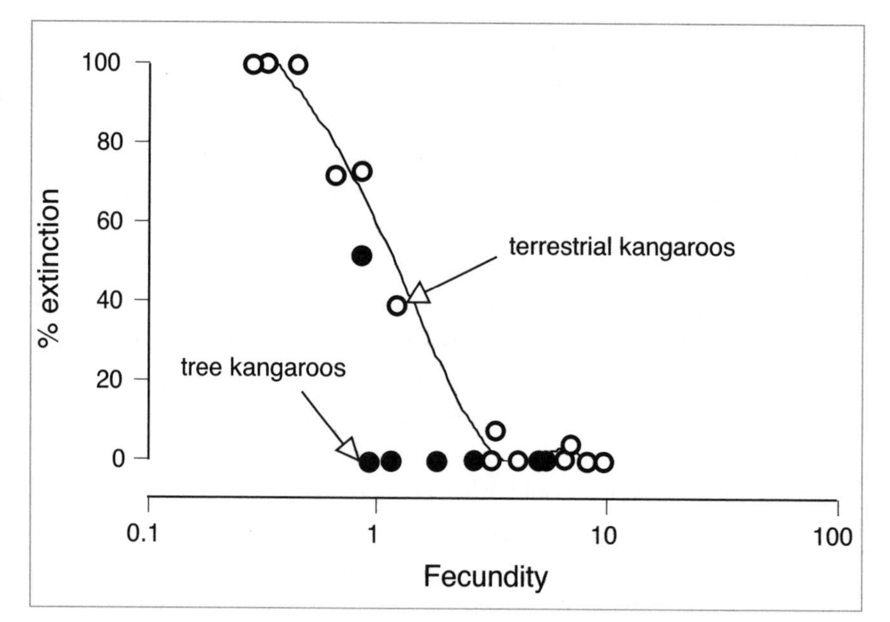

FIGURE 6.2. The relationship between the mean fecundity (offspring per female per year) in families of marsupials present in Australia during the late Pleistocene, and the percentage of species that went extinct. Families of terrestrial species (○) are distinguished from families of arboreal species (●); the Macropodidae (kangaroos) are split into terrestrial and arboreal sections.
Notes: The line is a polynomial fit to the points for terrestrial groups (by linear regression, $R^2 = .90$, $F_{1,11} = 100.46$, $p < .0001$).

Plates 17 Rock art from Arnhem Land depicting a thylacine. The painting shows a dog-like animal, but with the cloaca of a marsupial and the thick tail-base of a thylacine. The animal has a three-pronged spear lodged in its back, just behind the shoulder. (photograph by George Chaloupka)

Plate 18 Tree-dwelling mammals like the scaly-tailed possum *Wyulda squamicaudata* suffered less decline in recent years than ground-dwelling species of the same body size. (photograph by Euan Ritchie)

PLATE 19 Species like rock-wallabies that shelter in rock-piles and on rock faces have suffered less decline than species that live on open country. This is a Mareeba rock-wallaby *Petrogale mareeba.* (photograph by Michael Cermak)

PLATE 20 The burrowing bettong *Bettongia lesueur* has disappeared from mainland Australia and survives only on islands off the Western Australian coast. Mainland populations have recently been established in fox-free reserves. (photograph by Jiri Lochman)

species with dangerously low reproductive rates, but most of the slow-reproducing arboreal species survived. In particular, there was a lower extinction rate in tree kangaroos than among terrestrial kangaroos, even though tree kangaroos are demographically very sensitive to over-hunting.

There are also some hints that, among the extinct large terrestrial species, those from densely vegetated habitats survived longer than related species from open habitats. The kangaroo species for which we have the strongest evidence of survival past 46 kyr ago is *Simosthenurus occidentalis*, which may have persisted for 10 kyr or more after the arrival of people in southwestern Australia (Prideaux *et al.* 2000; Roberts *et al.* 2001b; Prideaux personal communication). The foot anatomy of *S. occidentalis* suggests that it used very uneven or hilly terrain, and it probably lived in relatively inaccessible and densely vegetated habitats (Bishop 1997; Prideaux 2004). Also, there are hints of late survival (from less securely dated contexts) of several large terrestrial kangaroos and a Diprotodontid from dense rainforest in New Guinea long after their open-country relatives went extinct in Australia (the species are *Maokopia ronaldi*, *Protemnodon hopei*, *P. nombe* and *P. tumbuna*; see Table 2.1 and Table 4.2 for details).

Interpreting the relationships of reproductive rate, habitat and habit (arboreal or terrestrial) of late Pleistocene species to their probability of extinction is not straightforward, because all three attributes are themselves related to body size. Reproductive rates decline with body size; the very largest mammals tend to occur in open habitats rather than in dense forest; and all of the larger species were ground dwelling. We already know that the extinctions were strongly size-selective, but this size-selectivity could have been caused in two ways: indirectly, because of the association of reproductive rate (and so on) with body size, or directly, because large species were more obvious to predators or were targeted because each individual provided a bigger reward for a hunter.

There are hints in the extinction pattern that reproductive rate had a strong bearing on extinction, independent of body size. Echidnas weighing 10 and 30 kilograms went extinct, while there were very few extinctions in that size range for kangaroos and wombats; this can be explained by the fact that echidnas have exceptionally low reproductive rates for their body size. Also, it is interesting that the larger echidnas survived in dense forest in New Guinea but species of about the same size disappeared from more open habitats in Australia.

I have done a larger analysis of the traits linked to extinction in megafauna species worldwide, including marsupials, which demonstrates

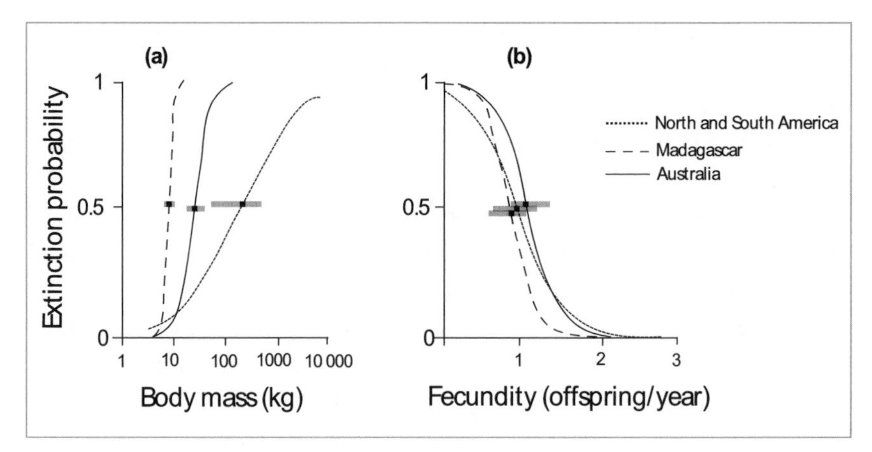

FIGURE 6.3. The relationship between probability of extinction and **(a)** body mass or **(b)** fecundity in late Pleistocene mammals from the Americas, Madagascar and Australia.
Notes: The lines are the fitted probability functions produced by logistic regression. Black squares and horizontal bars are the predicted values of mass or fecundity, with 95 per cent confidence intervals, at which the probability of extinction reached 50 per cent. From data in Johnson (2002).

that species with low reproductive rates were the most likely to go extinct in the late Quaternary, regardless of body size (Johnson 2002). This is illustrated in Figure 6.3, which compares the relationships of extinction probability following human arrival to body mass and reproductive rate in Australia, the Americas and Madagascar (the three major land masses first occupied by humans within the last 50 kyr). In each place the probability of extinction increased with body mass, but in quantitatively different ways. Mammals from Madagascar and Australia went extinct if larger than about 8 and 40 kilograms respectively, while in the Americas extinction was likely only for species weighing several hundred kilograms or more. However, the relationship of extinction risk to reproductive rate (which is closely related to body mass, but in different ways in different mammal groups) was remarkably consistent (Figure 6.3b).

A good generalisation is that all mammal species in which females produced less than one offspring per year went extinct after the arrival of people in a new land, whether they were quite small species like some marsupials and lemurs, or giants like mammoths and ground sloths. The only consistent exceptions to this rule are slow-reproducing survivors such as tapirs or long-beaked echidnas that live in densely vegetated habitats, or sloths or tree kangaroos that live in trees.

Population density might also have affected the risk of extinction from overhunting, because species that were originally present in low numbers would have been more strongly affected by a given off-take by hunters. Population density typically declines with body size, so an effect of population density on extinction would also have contributed to high extinction rates among the largest species. But just how population density affected extinction risk would have depended on the extent to which the hunters specialised on certain prey. If people did specialise on just a few species of prey, they would presumably have singled out those that were sufficiently abundant to provide a useful return for hunting effort, and if those species were driven to low numbers by overhunting they would have switched to more abundant prey. In this case low abundance might have provided some protection. But if people hunted any species that they encountered, low abundance would increase the risk of extinction. This is because the size of the population of hunters would be insensitive to the abundance of any particular prey species, and the hunters would continue to pose a threat to individuals of a rare species even as its population declined towards extinction.

Were the extinctions selective for slow-moving species? The movement styles of those extinct species for which we have reasonably complete skeletal material have been analysed by Murray (1991), who provided the following insights. Among the kangaroos, *Simosthenurus* species had long and sturdy hip bones, but relatively short lower leg segments. This probably fitted them better for short bursts of fast hopping than for maintaining speed over long distance. Their escape strategy probably involved short sprints into cover. *Procoptodon* had longer limbs and could have sustained higher speeds, but they were probably slow starters. The *Protemnodon* species had stocky hind limbs and short feet, and so probably benefited much less from the elastic recoil mechanism that makes bounding so energy-efficient at high speeds for living kangaroos. Like *Simosthenurus* they could probably accelerate quickly but at a high energy cost that limited them to short sprints.

The large extinct macropodines were probably quite similar to living *Macropus* species in their movements, but have not been analysed in detail. Diprotodontids had short limbs and little capacity for sprinting but 'were probably capable of a sustained, smooth and efficient, but comparatively slow, maximum speed gait' (Murray 1991). The limb proportions and joints of the extinct wombats, including the very largest species, were very similar to the living species and are consistent with burrowing. Like the living wombats, the extinct wombats were probably good short-distance sprinters but poor long-distance runners.

In summary, while it seems that there was a range of styles of locomotion among the extinct species and that some were capable of moving quickly over short distances, it could be that few of them were as well equipped for long-distance fast movement as the living kangaroos. Whether this means they were peculiarly vulnerable to human hunters depends a great deal on what tactics those hunters used – and these are, of course, unknown. But I suspect that extinction in the late Pleistocene was not simply a matter of 'the quick and the dead'.

THE POWER OF LATE PLEISTOCENE HUNTERS

The third test is much more difficult than the previous two, and it has not even been tried for climate change and fire in Australia. Martinez-Meyer *et al.* (2004) provided an example of how quantitative methods could be used to test a hypothesis of climate-driven extinction, in this case the end-Pleistocene extinctions of mammal megafauna in North America. They modelled the environmental niches, defined by temperature and rainfall, of eight extinct megafauna species, and used climate reconstructions to show how the habitable areas for these species changed through time. The analysis showed that suitable environments actually became more widespread as the animals declined to extinction. This is evidence that climate change did not contribute to their extinction.

The overhunting hypothesis is much more amenable to quantitative analysis using models. The interaction at its heart – people killing animals – can easily be made explicit in a formal model. Ecological theory provides the mathematical tools needed to simulate the dynamic interaction between populations of megafaunal prey and their human predators, and the model elements that describe the independent dynamics of the predator and prey populations – their maximum rates of increase and maximum densities, and functions specifying the effect of density on rate of increase – can be estimated with fair confidence.

The models must also specify the rate at which people might have hunted megafauna. Brook & Bowman (2004) did this using two parameters: hunting off-take (the number of prey individuals killed per person per year) when the megafauna population was large, and 'prey naivété'. The latter described how predation efficiency changed as density of the prey declined, and can be interpreted as the development of avoidance of hunters in response to the initial onslaught.

A simulation using such a model typically begins with a pristine environment with megafauna at carrying capacity, introduces an initially small human population, runs the model forward until a steady state is reached, then repeats the simulation many times with different parameter values to test the robustness of the final state of the model. Alroy (2001) modelled the impact of people on mammal megafauna in North America in this way and Holdaway & Jacomb (2000) did the same for people and moa in New Zealand, while Brook & Bowman (2004) created a more general family of models that provided results applicable to Australia.

All these studies concluded that extinction of large vertebrates was a consistent, almost inevitable consequence of human arrival, because the reproductive rates and densities of these large vertebrates were low relative to the potential population density of people and the predation rates they could have imposed. The models also concluded that the extinctions would have been completed quite quickly, typically around 700 years in Brook and Bowman's simulations. This is a long time on the scale of human (or megafaunal) lifetimes but, set against the 45 kyr span of human occupation of Australia, a decline that was spread over 700 years would appear to be instantaneous and would certainly not be resolvable by the coarse dating that is possible for events of this time-depth. It would look a lot like a blitzkrieg, whether or not it seemed like one at the time.

An earlier modelling exercise, tailored more specifically to Australia, concluded that very rapid extermination of megafauna would have required unrealistically high search efficiencies and hunting rates by people (Choquenot & Bowman 1998), but that study was framed on very short time scales and does not contradict the conclusion reached by Brook & Bowman's (2004) more general model.

These models are grossly simplified versions of reality that leave out many details of the interaction between people and megafauna, some of which assume overwhelming importance at the point where prey populations have been driven down to low abundance. One such detail is the motive for hunting. This is usually assumed to have been for subsistence, but there are other possibilities. For example, Brook & Bowman (2002) pointed out that if the purpose of hunting was symbolic or to gain prestige the hunting pressure might have been far higher than is normally assumed, and would have increased as megafauna became rare. This would have made extinction more likely.

But the most celebrated of these details is the assumption of prey naivety, which in this modelling context means that megafauna, initially easy to kill, did not become less easy to find and kill as their populations declined under hunting. Brook & Bowman (2002) showed that results of previous simulations depended heavily on an implicit assumption of prey naivety so defined; of their own simulations, the one that best reproduced the observed size-biased pattern of extinction assumed a high degree of prey naivety. If our hypothetical megafauna had become wary of people, the modelling suggests that this might have saved them. But it might also have increased their vulnerability by another route: maybe they became so wary that they avoided places used by people. If these places were waterholes and the prime grazing lands around them, the mere presence of people in a landscape would have reduced its carrying capacity for megafauna. This effect can also be incorporated in Brook & Bowman's (2004) model by including a suppression of the megafauna carrying capacity as a function of the presence of people, producing an extinction pattern as realistic as that produced by hunting alone. So, we could argue, if we chose, that avoidance of human hunters should have protected megafauna from extinction, or that it increased their risk of extinction, and we could illustrate each result with a plausible simulation. This is the point at which modelling ceases to be helpful.

However, the main problem with these models is the assumption that people ever hunted megafauna at the levels needed to have significant effects. Past presentations of the overhunting hypothesis, especially the highly coloured blitzkrieg version, give the impression that people killed megafauna in large numbers. They also imply that all age classes in the prey populations, including the very largest and most formidable individuals, were tackled by hunters. This is why the hypothesis calls up those images of 'man the mighty predator' and seems to require archaeological evidence of specialised hunting on a very large scale.

But the archaeological record shows almost nothing of this. Not only are there very few remains of megafauna with evidence of use by people (see Chapter 4), but there is no archaeological evidence from that time of weapons that might indicate specialised hunting of large mammals (see Chapters 3 and 8). It should be noted, however, that this has not led any archaeologist to conclude that early people never hunted megafauna. Mulvaney & Kamminga (1999) expressed this common sense view:

> While direct evidence is lacking, we assume that pursuit of at least the smaller megafauna was well within the capacity of Pleistocene hunters, given that people hunted animals as large as red kangaroo, crocodile, emu, seal and sea lion with equipment of very simple design, such as clubs and a range of spears including thrusting javelins.

Large animals might have been ambushed around waterholes, and a hunter who inflicted a wound in a brief surprise attack could then have tracked the stricken animal to exhaustion. It is easiest to imagine the very largest and most formidable species being hunted like this, but the smaller of the extinct species could have been killed outright with wooden spears, as were the largest kangaroos in recent times. The large reptiles and birds might have been hunted in similar ways, and no doubt their eggs were robbed from ground nests.

So, if we assume that there was some hunting of megafauna, the critical question is: how much would have been enough to drive species extinct? This is a question to which modelling can give a clear answer, as I will show with a demographic model of a *Diprotodon optatum* population. This model used demographic values inferred from relationships with body size established for living large marsupials (Johnson, unpublished). These predict a reproductive rate of 0.32 offspring per adult female per year, an age at maturity for females of 6 years and a maximum lifespan of 40 years.

I used these values to draw up an age-specific fecundity schedule for *D. optatum*, and then fitted a survivorship schedule that would equalise total births and deaths. The survivorship schedule had plausible features for a large mammal – a high survival rate for mid-age animals, intermediate survival rates for juveniles, and a steep decline in survival rates in the oldest age classes (Caughley 1977). I gave the population a maximum rate of increase (R_{max}) of 1.10, very close to the R_{max} measured for the white rhino by Owen-Smith (1999), meaning that if the population was reduced to low numbers it would tend to recover at a rate of 10 per cent per year. Density dependence of population growth was modelled assuming contest competition, and the population's dynamics were simulated in the program RAMAS Metapop (Akcakaya 2002). The simulations consisted of setting a carrying capacity of 1000, allowing the population's age distribution to stabilise at that size, then finding the size of the population at equilibrium with each of a series of simulated off-take rates. Simulations were run over 1500 years, about twice the period that emerged from Brook and Bowman's modelling as a typical time to extinction. I began by modelling very small off-takes by hunters, then increased these to find the point at which the population went extinct.

I found that the simulated population was able to absorb an off-take of up to 29 animals each year (that is, 2.9 per cent of the initial size of the population). The taking of 30 animals per year resulted in extinction after 520 years. *D. optatum* may have lived at population densities of around two animals per square kilometre. A population initially of 1000 animals would therefore occupy an area of about 500 square kilometres. Lourandos (1997) estimated the pre-European density of Aboriginal people in inland southwestern Victoria (in habitats that would have been occupied by *D. optatum* when people first reached them) at 0.3–0.4 people per square kilometre. At the lower end of this range, a 500 square kilometre area would have been occupied by about 150 people. The rate of hunting that wiped out the simulated population therefore represented the killing of only one animal every five years for each person – or, perhaps a bit more meaningfully, two animals per year taken by each group of ten people.

This first round of simulations assumed an unselective harvest: that is, animals young and old were killed in proportion to their relative abundance in the population. I repeated the simulation with the assumption that only juveniles (less than six years old) were killed by hunters. In this case the maximum sustainable off-take was 37 animals per year, with a rate of 38 causing extinction after 730 years. The unharvested population contained 414 juveniles at its stable age distribution, so the unsustainable harvest represented only 9 per cent of their initial number.

Populations of long-lived vertebrates are usually much more strongly suppressed by increased mortality of adults than of juveniles, because the naturally high survival of breeding adults means that each adult female has many opportunities to replace offspring that die young. But in this case the fact that adult females produce only single offspring that do not mature until six years old means that juveniles killed by hunters are replaced very slowly, and a small increase in mortality of juveniles can hold the recruitment rate below levels needed to replace natural mortality of breeding adults. As a result the maximum harvest of juveniles was only slightly greater than for an unselective harvest that included removal of breeding adults.

In this scenario we need not wonder how early Aborigines with their simple weapons could possibly have brought down an animal that weighed more than two tonnes, and that might well have learnt enough about people to have defended itself or run away. In species of large marsupials, sexual maturity is reached well before the completion of growth. A five-year-old diprotodon might have been only about half the size of an average full-grown animal, and it would have been a much easier prospect for a hunter.

These simulations specified an initial off-take in numbers of individuals killed per year, then continued to apply that number until the population either came into balance with it (as a result of the population's potential growth rate increasing as its density declined) or went extinct. In reality it would become increasingly difficult to take the same fixed number from the population as it became more sparse, and a more believable model would allow for a decline in the absolute off-take with declining prey density. Barry Brook and I have developed a more sophisticated version of the model that includes this effect, and showed that the main result is not changed (Brook & Johnson 2006). The demographic sensitivity to increased mortality of *D. optatum*, or any other animal like it, means that at any density it is very easy to overexploit the population and make it smaller still. The simulations suggest that this could have been done without ever killing a large number of animals within a short period.

So, the general answer to the question of how much hunting by people would have been enough to drive megafauna species extinct is that very low rates of killing, representing only small rates of mortality imposed on the hunted populations, could have done it. The problem of the lack of archaeological evidence of hunting can now be seen in a very different light. The simulations above show that the levels of hunting needed to drive populations of species like *D. optatum* extinct might have represented only a small investment in time by the hunters and made only a minor contribution to their food intake. In that case a sensible expectation would be that large-animal hunting should have contributed only a small proportion of the archaeological evidence of the early occupation of Australia, and therefore that direct evidence of hunting should be rare. This fact, together with the general rarity of bone in archaeological sites more than 40 kyr old, and the fact that even a very gradual extinction could have been completed within a thousand years, can explain the near-absence of direct archaeological evidence of killing of extinct species.

Also, because the additional mortality imposed by hunters may well have been small, the majority of individual animals would have died natural deaths even as their populations declined to extinction as a result of overhunting. This is an especially strong result of the simulation that restricted hunting to juveniles: in this case all adult deaths were due to causes other than hunting over the 700 year period during which hunting was driving the population extinct. Therefore the fact that the most recent remains of megafauna that we do have (all the best of which are from non-archaeological contexts) mostly show no sign of having been killed or used by people does not constitute evidence against overhunting as the cause of their extinction.

THE DWARFING TEST

Dwarfing of survivors was a process closely related to the extinction of the megafauna. It increased the power of that event by ensuring that only small-bodied animals survived. A successful hypothesis of the cause of extinction should, therefore, also be able to account for dwarfing. The climate change hypothesis cannot do this. Supporters of climate change might be able to argue that the stresses of the LGM selected for smaller body size, perhaps to minimise whole-body energy or water requirements. This would be a difficult argument to sustain, for the reason given above: large body size was probably an advantage in glacial climates. But there is a worse problem. On this argument, dwarfing should have gone into reverse as the LGM gave way to the present interglacial climate – if the LGM changed the giant *Macropus titan* into the miniature eastern grey kangaroo *M. giganteus,* the Holocene should have given *M. titan* back to us. The fact that this did not happen tells us that the pressure that caused dwarfing in the first place was maintained thereafter, not relaxed. The fire hypothesis does not provide any clear prediction that survivors of the habitat change caused by burning should have become smaller.

On the other hand, hunting should have caused dwarfing, and if hunting pressure on survivors was maintained (as it was) then the dwarfed animals should have stayed small. Hunting increases the mortality rate in wild populations. An adaptive response to this in a hunted species is to increase the reproductive rate and to shorten the generation time. Both can be achieved if females start breeding at earlier ages. Within the life history of an individual, growth and reproduction trade off against one another. An animal initially grows, then stops growing at a certain age and transfers its energy surplus into reproduction. Shifting this growth/reproduction switch to an earlier age results in more rapid population growth and a reduction in mature body mass. Hunting therefore transforms large slow-breeding mammals to small fast-living ones (if it does not drive them extinct).

The effect of hunting on body size is even more obvious if large individuals are selectively killed, but evolution to smaller size is a predictable result even if hunting is non-selective. Modelling of this process in large mammals shows that a 10 per cent proportional increase in mortality rates leads to a 21.3 per cent decrease in adult weight and a 9 per cent drop in age at maturity (Purvis 2001). The effect declines for species that were initially smaller (Purvis, personal communication), so it can account for the size-dependent pattern of dwarfing that occurred in kangaroos (Figure 2.7). The process has been seen in action in several fisheries, in which individuals have become genetically smaller (and therefore less valuable) under harvest (Law 2001).

AN ANSWER

The chronological evidence supports direct human impact, not climate change or fire, as the cause of the megafauna extinctions. This evidence is strong, but perhaps not yet overwhelming. We still have rather few hard dates for events in the last glacial cycle, and it could be that new dating studies will change our understanding both of when the megafauna went extinct and when people came to Australia. Anyway, as Chapter 4 showed, the date of 46 kyr ago for the extinction event is not yet accepted by everybody, and it might be many years before all doubt is removed over the timing of the extinctions and their relationship to human arrival.

The evidence from the selectivity of extinctions is much clearer. The extinction pattern is the one we would have expected had overhunting been the cause. Species that reproduced slowly, and that lived on the ground in open habitats where they were easily accessible to human hunters, went extinct; species with the opposite characteristics survived. On the other hand, the species that should have been most sensitive to the environmental changes of the last glacial cycle survived, while those that should have been most resilient to them went extinct. It may be true that the climate became more harsh and variable through the last 400 kyr, but there is no reason to think that this should have been particularly bad for very large mammals. There is no doubt over how to interpret this evidence: environmental change could not possibly have been the cause of the extinctions, but hunting by people might well have been.

Quantitative analyses using population models show that an invading human population could have caused the selective extinction of large vertebrates. These models make it clear that low rates of hunting, possibly extending no further than the occasional taking of juveniles in the largest species, could have resulted in megafauna extinctions on time scales of a few centuries. Therefore, to explain the extinctions there is no need to invoke a concentrated hunting effort on large mammals that would have required specialised weaponry and left a clear signal in the archaeological record. The killing of a big mammal may well have been a rare event in the daily lives of the first Aborigines. Equally, we can do without the idea that the megafauna were especially naive, the purpose of which was to explain how it could have been that large numbers of them were killed within short periods of time.

The overhunting hypothesis has in the past been reconciled with the lack of archaeological evidence for killing of large mammals by invoking the speed of the extinctions – that is, it all happened so quickly that there was little opportunity for archaeological evidence to be laid down. I take a very different view of this problem: the hunting of the megafauna is archaeologically invisible because it was never a major activity of the people concerned, and at any one time it affected only a small proportion of the individuals in the hunted species. In addition, it belongs in a period of prehistory that has yielded precious little evidence of any hunting and gathering activities, and in particular very little bone of any kind preserved in archaeological sites.

It is a fairly safe inference that the people who hunted these extinct species did not specialise in hunting large prey, and they probably had a broad-based foraging economy. This makes sense in the light of their rapid initial spread through Greater Australia (see Chapter 4), because a lack of ecological specialisation would have allowed them to move easily from one environment to another. The gradual disappearance of some of their prey would not greatly have disadvantaged them, because the fall in food supply could have been made good by slight increases in the intensity of use of other resources less susceptible to overharvest.

It might be that the details of the interaction between people and megafauna will never be known, but the scenario sketched out above is probably the one most consistent with the patchy evidence we have. This evidence all points to the extinctions as being a consequence of hunting by people. The other two major hypotheses fail badly: in fact, it is a struggle to find *any* fact that indicates an effect of climate change or fire on mammal populations in late Pleistocene Australia. Therefore, there is no case to invoke multiple causes of the extinction. Although there is no doubt that a wealth of biological detail, most of it unknowable, determined just how each of the extinct species was affected by human hunters, extinction by overhunting is both sufficient to account for what happened and well-supported by the three classes of evidence that must be used to test the merit of any hypothesis.

Certainly the process has been repeated many times since. Wherever humans have harvested from their environment, long-lived, slow-breeding and slow-maturing species, living in situations that guaranteed high exposure to people, have been the most likely to disappear. Consider one of the more recent examples. In the early 1960s albatrosses began to decline throughout the southern oceans. A well-studied population of the wandering albatross *Diomedia exulans* at Bird Island in the South Georgia group declined steadily by about 1 per cent per year from 1960 to 1995, and more serious declines happened elsewhere (Tuck *et al.* 2001). The cause, it is now clear, was the incidental catching ('bycatch') of birds on hooks set by commercial boats fishing for (mainly) southern bluefin tuna. The birds would occasionally catch the baited hooks as they were cast on long lines, and were then dragged underwater.

Tuck *et al.* estimated that the annual bycatch for Bird Island averaged about 40 adult females per year between 1960 and 1985. This was small in comparison to the population's initial size of 3000 females and would have been barely noticeable to the tuna fishers, but it accounted for the observed declines. In the absence of changes to fishing practice, long-line fishing has the potential to drive this and other large seabirds extinct, and may yet do so, because the open ocean provides no places where large seabirds can live out their lives without encountering fishing boats.

If one accepts this illustration – in which a small increase in mortality from a single source, imposed on populations of very widespread birds with naturally high adult survival but very low fecundity, could eventually push them to extinction – it becomes easier to imagine that the same kind of thing happened to the marsupial megafauna.

7

The ecological aftermath

THE LATE PLEISTOCENE megafauna of Australia were a highly diverse collection of species, but one ecological category predominated among them: large browsing herbivores (see Chapter 2 and Plate 11). Through most of the Pleistocene, shrubs and small trees across inland Australia were browsed by upwards of 35 species of large kangaroos and diprotodontids. These megabrowsers probably made up the bulk of the mammal biomass at any one site.

This is illustrated in Figure 7.1, which reconstructs the distribution of biomass in relation to body size among the species of herbivorous mammals that lived around Cathedral Cave in southeastern South Australia. This cave has the best-preserved and most thoroughly documented mammal fauna of any middle Pleistocene site in Australia. Deposits dated to about 250 kyr ago, when the climate was evidently similar to today's, contain an assemblage of 37 herbivorous mammals, of which 11 are now extinct (Prideaux, personal communication). The non-extinct species represent a mammal herbivore community similar to that of recent times, ranging from pygmy possums and small rodents up to the common wombat and eastern grey kangaroo. The extinct species were a large koala, a further nine kangaroos,

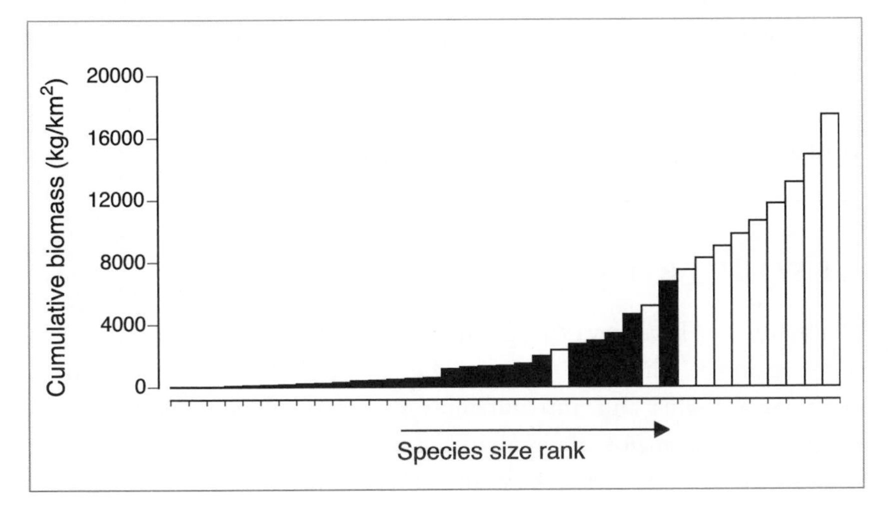

FIGURE 7.1 The cumulative biomass of herbivorous mammals in the middle-Pleistocene community around Cathedral Cave near Naracoorte, South Australia.
Notes: The species present at the site were determined by Gavin Prideaux (personal communication). I estimated the biomass of each species by multiplying its body mass by its population density. For living species (shaded bars), densities were measured in population studies (where no such data were available for a species, I used a figure from a closely related species), and for extinct species (unshaded bars), densities were inferred from the relationship of density to body mass among living large terrestrial herbivores (Johnson unpublished).

and the Diprotodontid *Zygomaturus trilobus*, all of which were browsers. Although the larger species probably had low population densities, their great size meant that they made up most of the mammal biomass around the site – the extinct herbivores, even though they represented only 30 per cent of the species in the community, contributed 67 per cent of the total biomass. They would have accounted for a correspondingly large proportion of the plant material eaten by mammals.

We know that large mammal herbivores can have huge impacts on vegetation. This is now most obvious in Africa, the only continent that still has something close to its full complement of Pleistocene megafauna. Leaving aside major effects of climate and intensive human settlement, herbivory by very large mammals is the dominant factor controlling vegetation over large areas of Africa (Owen-Smith 1988). The direct effects of African megaherbivores on plants extend from suppression of growth by heavy browsing to destruction of whole plants. There may also be indirect effects. Plant species that are heavily browsed are put at a disadvantage in interspecific competition; shrubs and trees kept small by repeated cropping are more vulnerable to fire; and herbivore damage to plant tissue may expose it to attack by parasites and pathogens. In many parts of Africa elephants and black rhinoceroses maintain very open savanna where there would otherwise be dense tall woodland. There is no doubt that if large mammalian herbivores suddenly disappeared from Africa the vegetation would change quickly and dramatically. Did something like this happen in Australia?

In this chapter I sift through the evidence to find hints of the kinds of changes that did follow the extinctions, but I begin with a review of some strange features of living plant species that make sense only as evolutionary responses to the presence in the landscape of very large vertebrates. There are many examples of such anachronisms – they provide the most direct evidence of the roles once played by megafauna in the ecology of Australian plants.

PHOTOGRAPHS OF GHOSTS

OBSOLETE DEFENCES

Herbivory can have devastating effects on plants, so plants defend themselves against it. These defences may be chemical or structural. Chemical defences – toxins and anti-nutrients stored in plant tissue – can be effective against small herbivores, including invertebrates, as well as against large mammals. Of interest here are structural defences that are specially directed against larger vertebrates. Spines – including thorns, prickles and rigid pointy branchlets – are the most important of these (Grubb 1992). Because mammals have tender lips and mouths they are sensitive to spines. Large browsing mammals will eat spiny plants, but they feed gingerly and this reduces their feeding rate and therefore their impact on the plant (Belovsky *et al.* 1991; Cooper & Owen-Smith 1986). The effect of spines is often increased by other plant structural traits such as small leaves, which make it difficult for the herbivore to take large mouthfuls of foliage, and a densely branched thicket-forming growth habit, which acts as a kind of cage to protect the main stems. Plate 12 shows an example of such traits in a small tree from inland Australia, the scrub leopardwood *Flindersia dissosperma.*

Spininess is common in many species of shrubs and small trees in Australia, especially in the genera *Hakea, Solanum* and *Acacia.* No living mammals eat *Hakea,* and consumption of *Acacia* leaves by terrestrial herbivorous marsupials is rare (Irlbeck & Hume 2003). The spiniest species of *Solanum* occur in the dry inland of Australia, where there are few browsing marsupials (Symon 1986). Low (1998) and Barker *et al.* (1999) argued that the spines on these and other plants evolved as defences against large browsers that are now extinct.

We can see the imprint of extinct herbivores on living plants most clearly in cases where the deployment of spines and other structural defences matches the stature of Pleistocene megafauna species. A good example is waddywood *Acacia peuce,* a large desert tree from the margins of the Simpson Desert. When young, *A. peuce* is densely branched and has rigid sharply pointed and outwardly directed phyllodes. This spiky form is maintained until the plant reaches a height of two to three metres, when it produces a main central stem, assumes an open growth form with relatively soft phyllodes and a weeping habit, and sheds its spiny lower growth (Jon Luly, personal communication; Plates 13 and 14). (*A. peuce* actually passes through three distinct growth forms: very small seedlings are grass-like and have foliage which, though soft, has a very strong smell that is rather like stale urine and induces headache (Tony Boland, personal communication).

Similar height dimorphisms are found in other tree species of inland Australia, such as leopardwood *Flindersia maculosa,* scrub leopardwood, wild lime *Citrus glauca,* and several species of *Capparis,* especially wild orange *Capparis mitchellii* and narrow-leafed bumble *C. loranthifolia* (Anderson 1993; Cunningham *et al.* 1981; Jacobs 1965). Ironwood *Acacia estrophiolata* changes from an intricately branched tangled shrub with small phyllodes to an erect tall tree with long pendulous phyllodes (Des Nelson, personal communication). The young phyllodes are not spiny, but they have a bitter taste that is lost as the plant assumes its mature form (Peter Latz, personal communication).

In many more tree species there is a less pronounced change in branching architecture with growth, but small plants are spiny while mature plants are not. Good examples are found in the genus *Bursaria*, particularly *B. incana* and *B. tenuifolia* (Cayzer *et al.* 1999), in *Pittosporum lancifolium* and *P. spinescens* (Cayzer *et al.* 2000), in some *Solanum* (*S. macoorai* and *S. viride*, Symon 1981), and in other acacias. In *A. pickardii* juvenile phyllodes are slender and rigid, aligned closely with the stem axis, terminate in a sharp point, and are further protected by an outwardly directed basal spine (Jon Luly, personal communication; Plate 13). As the plant matures the phyllodes expand, soften and grow away from the stem, and the spines are lost.

In these examples the change from defended to undefended foliage occurs quite consistently at around two or three metres in height, and does so even in specimens that are not subject to browsing by recently introduced livestock. Two to three metres would have been the approximate browse height for many of the late Pleistocene large herbivores, such as the giant kangaroos and diprotodons. Many of these plants produce mature foliage that is highly palatable to livestock. *Capparis mitchellii*, for example, is regarded as an excellent fodder tree and is lopped during droughts for sheep and cattle (Cunningham *et al.* 1981). Probably the growth strategies of these trees evolved to protect foliage within reach of the now-extinct large terrestrial herbivores.

A similar case has been described in New Zealand, where a high proportion of shrubs and trees have 'divaricate' growth, characterised by small leaves on thin twigs that branch at wide angles to form a tangled structure within which the leaves are inaccessible. Divaricate growth has long been seen as a structural defence against browsing by extinct moa. This interpretation was given weight by Bond *et al.*'s (2004) demonstration that the divaricate form reduces the rate at which modern analogues of the moa (emus and ostriches) are able to pluck leaves, but fails to protect the plant against mammal browsers recently introduced to New Zealand (goats can bite through the thin stems). Divaricate shrubs that grow to tree height switch to a non-divaricate form at about three metres. Some Australian plants also have divaricate growth: bullwaddy *Macropteranthes keckwickii* is an example (Murray & Vickers-Rich 2004).

The existence of such elaborate defences against large browsers has two implications. First, it tells us that the extinct large herbivores must have put heavy browsing pressure on plants in Pleistocene Australia. From the plant's

PLATE 21 Desert rat-kangaroo *Caloprymnus campestris*, from John Gould's *Mammals of Australia* (1863).

PLATE 22 Broad-faced potoroo *Potorous platyops*, from John Gould's *Mammals of Australia* (1863).

PLATE 23 Eastern hare-wallaby *Lagorchestes leporides*, from John Gould's *Mammals of Australia* (1863).

PLATE 24 Crescent nailtail wallaby *Onychogalea lunata*, from John Gould's *Mammals of Australia* (1863).

point of view, producing spines and defensive tangles of stems and keeping leaves small is costly, because tissue that might have been used to build photosynthetic capacity is instead deployed in defence. The fact that plants paid this price shows that the browsing pressure they were defending themselves against was high. Second, it is interesting that plants with these defences seem to be particularly common in two types of vegetation: open acacia-dominated woodlands in arid regions and (most of all) dry rainforest.

Dry rainforest – sometimes referred to as vine thicket, softwood scrub or, occasionally, bastard scrub – is a very distinct forest type that occurs across much of the north of Australia. It is a closed-canopy forest, complex and multi-layered (see Plate 15). Many of its plant species are related to wet rainforest lineages, but it typically occurs in regions with intermediate or highly seasonal rainfall. The trees are often deciduous in the dry season, but the dense canopy that renews in the wet season suppresses growth of grasses and keeps the forest floor open. Dry rainforest now has a patchy distribution and is restricted to sites where it is protected from fire. The prevalence in this vegetation type of plants adapted to large browsers suggests that it was a significant habitat for many of the extinct megafauna of northern Australia, and was once much more widespread than it now is. It is a fair bet that until as recently as the LGM much of northern Australia that is now open eucalypt savanna was still covered by dry rainforest.

BROKEN PARTNERSHIPS

An odd fact pointed out by Janzen & Martin (1982) is that many trees in central America produce crops of large fruit that fall to the ground and rot beneath the parent tree. The fruit is eaten by few or no animals, and the tree receives none of the benefits of seed dispersal that are enjoyed by plants whose fruit is eaten. Plants produce edible fruit for one reason only: to have their seeds dispersed away from the parent plant by an animal that eats the pulp and discards the seeds, or swallows the seeds whole and defecates them onto the ground. A plant that produces fruit that no animal eats is a puzzle. Janzen and Martin's solution to the puzzle was that these fruits had evolved to be eaten, and their seeds dispersed, by the extinct megafauna of central America, especially gomphotheres (*Cuvieronius*, relatives of elephants and mammoths) and Pleistocene horses. Consistent with this, some of the rotting fruit are quite similar to African fruits still eaten and dispersed by forest elephants, while others are now eaten happily by modern horses.

Janzen and Martin proposed features of a 'megafaunal dispersal syndrome', characterising plants that co-evolved with Pleistocene large mammals as seed dispersers. The description of this syndrome was refined and extended by Barlow (2000). 'Megafauna fruit' are large, fleshy and indehiscent, and they often have a tough outer rind. They have a nutrient-rich pulp; they set rather low on the tree, or drop to the ground just as they ripen; they are often dull brown, or orange or yellow, and aromatic when ripe; the seeds are often large and, if so, they may be protected from crushing by a tough endocarp; if the seeds are small they are numerous and are distributed through a very soft pulp, and they may also be unpalatable or laced with toxins to deter chewing.

Trees that produce fruits of these kinds and that are now bereft of their ecological partners are often patchily distributed, and they may especially be concentrated in lowlands and on flood plains, presumably because they are now predominantly dispersed by gravity or water – seeds can escape the vicinity of the parent tree only if the fruit rolls downhill or is carried away by a stream. Such trees are often able to regenerate vegetatively, by aggressive suckering. This can allow them to persist locally but it provides little capacity for populations to spread beyond the patches to which they are now restricted. There is one important class of exceptions to this distribution pattern: plants with fruits that are eaten and transported by people and have thereby become very widespread. Many of the fruits that we eat provide good matches to the megafaunal syndrome. We may have the mammal megafauna to thank for pawpaws, avocadoes, citrus, and maybe even apples and pears (Barlow 2000).

Australia also has trees that produce fruit matching aspects of the megafauna syndrome identified by Janzen, Martin and Barlow. Such trees are especially common in the dry rainforests of northern Australia, where 'many kinds of fruits and large seeds are left to rot in the trees or to accumulate on the ground below where they dry out, rot, or are consumed by insects' (Murray & Vickers-Rich 2004). These include some of the species discussed above that have structural adaptations to megafauna browsing, such as wild orange *Capparis mitchellii* and wild pomegranate *C. canescens* (see Plate 16). Murray and Vickers-Rich list many other examples, most of which they consider to have been eaten and dispersed by the extinct giant bird *Genyornis newtoni* and the more ancient mihirungs. A few of these fruits, such as emu apples (*Owenia*), continue to be eaten by emus, but Murray and Vickers-Rich suggest that emus are not tall or large enough to be effective dispersers of their seeds.

We do not know how much fruit might have been eaten by the marsupial megafauna. One of the extinct kangaroos, *Simosthenurus maddocki* (see Chapter 2) may have been a specialist frugivore, and it is possible that some other extinct kangaroos occasionally ate fruit, as do some tree kangaroos today. Captive wombats will eat fruit, so it might be possible that the larger extinct members of the wombat–diprotodon lineage ate some fruit along with the grass and browse to which their teeth are most obviously suited. The tree species that produce 'megafauna fruit' are concentrated in northern Australia, but at present our knowledge of Pleistocene megafauna is restricted almost completely to southern Australia. Perhaps there were northern megafauna species, as yet undiscovered, that were more frugivorous than their southern counterparts.

One other plant species deserves a special mention. The seeds of the rainforest tree *Idiospermum australiense* (or 'idiot fruit') weigh up to 225 grams and are the largest of any Australian plant (Edwards *et al.* 2001). The species is restricted to small areas in the Daintree lowlands and the foothills of Mount Bellenden Ker and Mount Bartle Frere in the wet tropics of northern Queensland. This extreme geographic restriction is probably due to the fact that the seeds are dispersed only by gravity and water, and so are concentrated at low elevations and beside streams, where the plant may be locally common. Graeme Harrington (personal communication) used translocation experiments to show that *I. australiense* is capable of establishing in upland rainforests, but there is currently no native animal in these forests capable of moving the huge seeds uphill.

The seeds are starchy and might well be nutritious, but they contain toxins and they do not have a surrounding pulp or a persistent endocarp. They are composed of from two to six readily separable cotyledons and have the unusual property that single cotyledons are capable of producing seedlings (Edwards *et al.* 2001); probably, fragments of cotyledons would themselves germinate to form viable seedlings (Will Edwards, personal communication). Perhaps the missing seed disperser for this species was a large-jawed vertebrate able to tolerate the seed toxins, that moved through the forest picking up whole seeds and, as it crushed them to break them into smaller chunks that it could swallow, dropped fragments from its mouth to the forest floor. What this animal might have been, and when it disappeared, is anybody's guess.

PLANT RESPONSES TO MEGAFAUNA EXTINCTION

The examples given above of defensive growth forms in Australian plants tell us that the extinct giant marsupials had an important role in (quite literally) shaping the vegetation of this continent. The removal of those herbivores should therefore have produced a large response in the vegetation. Below, I consider three aspects of this response: effects on the structure of vegetation, on fire regimes, and on the distribution and abundance of particular plant species.

VEGETATION STRUCTURE

With the removal of large herbivores, plants that had been suppressed by heavy cropping might have survived better, grown larger and become more widespread. Given the predominance of large browsers among the megafauna, these responses should have been strongest in shrubs and small trees. So, the density of shrubs and trees, and the total biomass of vegetation, should have increased.

Some pollen histories do show an increase in low woody vegetation at around the time the megafauna disappeared. The clearest case is the GC17 marine core, which records an increase in chenopod shrublands after about 45 kyr in the arid northwest of Australia (van der Kaars & De Deckker 2002). The Lake Wangoom and E55-6 records from southern Victoria also show increases in shrubs (particularly 'Asteraceae B'). These might have happened around 50 kyr, but dating of the middle part of the last glacial cycle in these records is uncertain (Harle *et al.* 2002). The Lake George pollen record suggests that chenopods, myrtaceous shrubs and Asteraceae were more common in the last glacial cycle than in the previous one, but the timing of the increase in the last cycle is very uncertain (Singh & Geissler 1985). Lastly, Luly (2001b) showed that for a period after about 50 kyr ago *Callitris* woodlands were more extensive in the Lake Eyre region than at a climatically similar time around 100 kyr earlier, and he speculated that the difference was due to the absence of large browsers in the later period (as noted in Chapter 2, there are hints that diprotodons may have browsed *Callitris*). Most of these records also show grasses increasing in the latter part of the last glacial cycle, but they did so somewhat later than shrubs and trees, in the immediate approach to the LGM when the climate was at its most arid. The exception to this is Lake George, where grasses appear to have declined through the whole of the last cycle.

If there was an increase in shrubs and small trees relative to grasses at 40–50 kyr ago, this might also be detected in changes in the carbon isotope composition of vegetation. Miller *et al.* (2005) measured carbon isotope ratios in dated eggshells of emus and used them to reconstruct changes in the C3/C4 composition of vegetation in the Lake Eyre region over the last 140 kyr. Living emus have broad diets that include both C3 and C4 plants, and the concentration of ^{13}C in eggshell reflects the balance of these plants in the diet. There was a sharp decline in ^{13}C values of emu eggshell between 50 and 45 kyr ago, suggesting that tropical grasses were replaced by shrubs or temperate grasses. The nature of the change was that, initially, ^{13}C concentrations in eggshells varied over a wide range – some birds ate predominantly C3 plants and others predominantly C4 plants. This variability could have been due to fluctuations in composition of the vegetation, caused perhaps by occasional strong incursions of the northern monsoon. After 45 kyr ago ^{13}C concentrations were far less variable, and consistent with a diet composed almost purely of C3 plants. Carbon isotope analysis of wombat teeth from southern South Australia and western New South Wales revealed a similar shift from a diet of C4 to C3 plants at around 50 kyr ago (Miller *et al.* 2005). This is good evidence for a widespread change in the composition of the ground vegetation through the dry inland of southern Australia at about 45 kyr ago, and this change probably consisted of an increase in shrubby vegetation and small trees.

These data are especially interesting because, as explained in Chapter 5, declines in atmospheric CO_2 and rainfall through the last glacial cycle should have produced an increase in C4 over C3 plants, not the reverse. A shortage of atmospheric CO_2 reduces the rate of C3 photosynthesis, but the C4 photosynthetic pathway includes a mechanism for concentrating CO_2. C4 plants are therefore resilient to declines in atmospheric CO_2. This also protects C4 plants from moisture stress, because they can reduce the extent to which they open their stomates to take in CO_2 and thereby limit evaporative water loss without compromising photosynthetic capacity (Ehleringer & Monson 1993). The competitive advantage enjoyed by C4 plants in low-CO_2 and dry climates explains why C4 vegetation expanded globally from about 8 million years ago (Cerling *et al.* 1997). Carbon isotope studies of buried soil organic matter or herbivore teeth from eastern equatorial Africa and southern North America agree in showing increases in C4 plants in the approach to the LGM (Johnson *et al.* 1999).

Why did the trend revealed by the emu and wombat data go in the opposite direction in Australia? Temperature also plays a part in determining the balance of C3 and C4 plants in vegetation because, as C4 photosynthesis is strongly limited at low environmental temperatures. Johnson *et al.* (1999) suggested that the low temperature of the LGM might explain why C4 plants did not increase in the Lake Eyre basin. However, it does not seem that a fall in temperature can explain the shift from C4 to C3 plants at 50 kyr, because at that time temperatures were moderate and stable. If temperature had contributed, the change should have begun at 30–35 kyr ago, when temperatures dropped to the low levels of the LGM. Perhaps the vegetation change indicated by this diet shift consisted of an increase in C3 shrubs that had previously been kept in check by megafauna browsing (Johnson 2005a).

To summarise: it is possible to step through the palaeoecological data on the second half of the last glacial cycle and find evidence for vegetation change caused by the megafauna extinctions – but this is rather a struggle. One problem is that very few pollen records are well dated for the critical middle part of the cycle, and fewer still take in the last two cycles. A good test of the effect of megafauna browsing on vegetation would be to compare vegetation changes in the second last cycle, through which large browsers were continually present, with the last, in which they disappeared. Because the last two cycles were climatically similar this comparison could perhaps separate the effects of climate change and megafauna browsing on vegetation.

At present the pollen data are not strong enough either to decisively show or to rule out an effect of the extinctions on Australian vegetation. The studies on isotopic signatures preserved in emu eggshell and wombat teeth are strongly suggestive of such an effect, but they would be more powerful if they too could be extended into the previous glacial cycle.

FIRE

If there was an increase in low woody vegetation following the megafauna extinctions it may well have caused changes to fire regimes. Flannery (1990, 1994) proposed that this did happen. His ideas on the topic are important because they link the megafauna extinctions to Aboriginal land management and to present-day conservation of mammal biodiversity. He argued that before they went extinct the large marsupials had held vegetation in check over most of inland Australia; with their demise there was a build-up of plant biomass, which created the potential for massive and destructive wildfires. Aboriginal people responded to this raised fire potential by developing a system of low-intensity patch burning that reduced fuel loads and effectively replaced the regulation of vegetation that had been provided by the megaherbivores. It follows that the cessation of traditional burning practices now will result in the conversion of plant communities to states that did not exist in either the Holocene or the Pleistocene, and could well cause extinctions of other species that depend on the habitat conditions that were created first by megafauna and then by Aboriginal burning.

If Flannery's model were correct, we should expect to see the following in long-term vegetation records. Before the extinctions, charcoal should have been relatively rare, because megafauna browsing left too little fuel to support hot or frequent fires. Very soon after, there should have been a vegetation change marked by the recovery of plants that had been heavily browsed. This vegetation change might have been associated with a charcoal spike, as hot fires burned through a continent suddenly made dangerously flammable; but charcoal levels should then have stabilised at an intermediate level, reflecting recurrent low-intensity fires set by people in landscapes with lowered fuel loads. This latter condition should have prevailed until traditional burning practices broke down as a result of the arrival of Europeans.

The unprecedented increase in charcoal required by this model does appear in the localised vegetation record of Lynchs Crater, but not in the larger regional history for northeastern Queensland depicted in the ODP 820 core, nor in any other of the published vegetation histories for late Pleistocene Australia. As described in Chapter 5 these histories show a variety of patterns of change in fire through time, with increases in charcoal before the megafauna extinctions just as common as increases afterwards. If anything, charcoal levels were unusually stable in the period from 50 to 40 kyr ago. Further, those pollen records with the best evidence for increased vegetation biomass after the megafauna extinction actually provide the *least* support for an associated increase in burning. In the GC17 record, charcoal values are highest between 75 and 80 kyr ago, remain moderately high until about 45 kyr ago, then drop to low levels for the remainder of the record. The Lake Wangoom and E55-6 records agree in having charcoal peaks in the early part of the glacial cycle, well before the megafauna extinctions, and showing low charcoal afterwards (until the end of the LGM at Lake Wangoom).

And while the events at Lynchs Crater might be taken as supporting Flannery's model, they actually provide a very poor fit. Here, the increase in fire was associated with a shift from rainforest to sclerophyll vegetation. Rainforest has very little leaf biomass within three metres of the ground, so it is difficult to imagine how a reduction of terrestrial browsing in rainforest could have raised its fire potential so much that it burned away and was replaced by eucalypt forest. One could argue that an increase in burning, triggered by megafauna extinctions, in the drier vegetation of the surrounding environment might have resulted in the penetration of rainforests by fire and brought about their eventual destruction. The problem with this argument is that the ODP 820 record shows that the regional pattern of burning through the last glacial cycle was not unusual: there had been a very similar sequence of events in the preceding cycle, which had neither people nor megafauna extinction.

This is all quite a surprise. Flannery's proposal might have been highly speculative, but it was very reasonable speculation. Although the evidence of vegetation change following the megafauna extinctions is patchy, there is enough of it to suggest that there were significant increases in fuel loads, and to raise an expectation that burning should have increased as a result. But the charcoal evidence points either to no widespread change, or in some cases to a reduction in fire in the aftermath of the megafauna extinctions. Why?

One possibility is that the kind of shrublands that expanded after the megafauna went extinct were not fire-promoting. In chenopod shrubland, for example, wildfires are rare (Leigh 1994). Apart from the fact that the bushes themselves have low flammability because the leaves are somewhat succulent and have a high salt content, fire is impeded because individual plants are typically surrounded by a halo of bare ground, where grasses and herbs are excluded by root competition and shading. In shrublands the shrubs themselves may contribute most of the flammable biomass, but the grass and herb layer provides the continuity that enables fire to spread from one shrub to another. The reticulate pattern of strips of bare ground that develops in mature chenopod shrubland acts as an effective system of firebreaks. It may have been even more effective under the climate conditions of the last glacial cycle because the large shrubs would have been more widely spaced as a result of low rainfall and low atmospheric CO_2, which would also have intensified the suppression by shading of plant growth beneath them. Added to this, some moderately large grazers – the wombats, and what we refer to as the 'large' kangaroos – survived the megafauna extinctions, and their grazing might have further suppressed grass establishment between the shrubs.

If chenopod shrublands increased in the aftermath of the megafauna extinctions, the effect might have been as if someone had thrown a fire blanket over vast areas of inland Australia. The same might have been true to some extent for other shrubland types. If this view turns out to be correct it could well reverse Flannery's interpretation of the development of landscape burning by Aboriginal people. It may be that the spread of fire-resistant habitats following the megafauna extinction made fire a less useful tool for Aboriginal people in the aftermath of the megafauna extinctions than it was later to become, and delayed the advent of 'fire-stick farming' by many thousands of years.

PLANT DECLINES AND EXTINCTIONS

The examples given earlier of plant species with growth forms adapted to resist browsing by large mammals, and with fruits adapted for seed dispersal by giant vertebrates, identify species that were especially strongly affected by megafauna. Such species might have undergone correspondingly dramatic changes in their distribution and abundance as a result of the extinctions. Unfortunately the fossil record for the late Pleistocene contains very little information that might reveal such changes for particular plant species, because pollen is generally difficult to identify below family level. But there are hints in the fossil record, and in the current distributions of some plants, that point to some major changes.

Considering first the plants with defensive growth forms, there could have been two very different effects of the loss of large browsers. On one hand, the fact that these species evolved such elaborate defences suggests they had been kept under especially heavy browsing pressure. The relaxation of this pressure may have allowed them to become more widespread and abundant. Alternatively, these species would have faced the problem that their inheritance of an energetically costly growth strategy that made sense in an environment that had many large browsers provided no benefit once those browsers had gone. In the post-megafauna world they might have begun to lose out in competition with species that did not invest in such elaborate and costly protection against browsing, and were therefore more efficient in capturing resources. Two possible results might be that the 'megafauna plants' declined to rarity and perhaps extinction, or that they underwent an evolutionary loss of their obsolete defences. These hypotheses have not yet been systematically evaluated and we cannot say which is most often correct, but the following observations are interesting and relevant to the problem.

First, some of the trees with the most extreme anti-browsing defences are now rare. *Acacia peuce*, for example, occurs only in three small disjunct populations on the margins of the Simpson Desert. It is found in monospecific stands, suggesting perhaps that it now persists only in sites too harsh for other species to establish and out-compete it. Michelle Waycott and Jon Luly (personal communication) have begun to investigate the genetic structure of the species, and their preliminary results show that the small surviving populations are genetically similar and retain quite high levels of genetic variance. This suggests that the current distribution is the result of the decline of a species that was until recently continuously distributed over a much wider area, and is consistent with this decline having begun after the megafauna extinctions.

Second, although the examples of anti-browsing adaptations in species such as *A. peuce* are dramatic in themselves, it is striking that Australian woodlands have fewer plants with these kinds of defences than places that still have large native browsers, or that lost them relatively recently. In African scrublands the majority of shrubs and trees have obvious structural defences against large browsers, but many woodland shrubs in Australia lack such defences. This might suggest that the evolutionary impact on plants of browsing mammals was never as great in Australia as it has clearly been in Africa. But maybe the difference is better explained as a result of plant extinction and evolution over the 45 kyr that have passed since the loss of the marsupial megafauna. Perhaps many plant species that thrived along with the megafauna, having evolved effective strategies to reduce the impact of browsing, followed the megafauna into extinction.

Looking ahead a few centuries it is easy to imagine that a plant like *Acacia peuce* may be added to a long list, for the most part hidden from us, of such plant extinctions. Many other species may have persisted but lost their spines and specialised growth forms. It is possible that this could happen quickly, as shown by the example of *A. nilotica*, an African species introduced into north Queensland. In its native range *A. nilotica* has long savage thorns, but these have been all but lost in some Australian populations (William Foley, personal communication). Peter Latz (personal communication) notes that in the Northern Territory the native species *A. victoriae* frequently lacks spines in many populations, but it is uniformly spiny in areas where it is heavily browsed by camels. In this respect camels may have replaced the marsupial megafauna and prompted the re-evolution (or at least the re-expression) of defensive structures.

The effect of the loss of a seed disperser is much easier to predict in principle. It should result in a reduction of the plant species' geographic range and genetic diversity, especially in species that have little capacity for vegetative reproduction. However, some of the species with 'megafauna fruit' were important in the diet of Aborigines, and perhaps for this reason they remain moderately widespread – although Murray & Vickers-Rich (2004) suggested that *Owenia* species are now in decline in the Northern Territory, partly because of insufficient seed movement either by people or animals. Nonetheless it might be worth reconsidering in this light one of the very few documented cases of plant extinction in the late Pleistocene, the disappearance of the rainforest conifer *Dacrydium*. This plant's last known occurrence in Australia was at Lynchs Crater, and it vanished just before the LGM as the rainforests were replaced by sclerophyll woodlands. The species in question was probably *D. guillauminii*, which still survives in New Caledonia where it is restricted to the vicinity of streams. Given the drying of Lynchs Crater at the LGM it is no surprise that it did not persist there, but it is a reasonable assumption that populations migrated to the wetter coastal lowlands, as presumably they had done in previous glacial maxima. The real question is this. Why did the species not recover after the LGM? Perhaps increased landscape burning was the reason. But as *D. guillauminii* is probably dispersed by large birds, could the absence of giant birds have been part of the reason for its failure to repopulate its former range?

CONCLUSIONS

We still have a great deal to learn about the role of the megafauna extinction in ecosystem change in Australia during the late Pleistocene. There is not much doubt that browsing by large mammals was a potent ecological factor through most of the Pleistocene. Our knowledge of late Pleistocene vegetation history is not yet good enough that we can be certain of how the vegetation changed when that browsing pressure was lifted. However, a reasonable working hypothesis, supported by some evidence, is that there was an increase in the density and extent of shrublands and low woodlands throughout much of inland Australia. It does seem quite clear, however, that this change did not lead to an increase in burning. If land management by Aboriginal people was influenced by the megafauna extinctions, the effect may well have been to reduce their use of fire rather than to promote it.

Part II

THE LATE PREHISTORIC PERIOD

[10 000 to 200 years ago]

Long ago, he used to light hunting fires
He used to light hunting fires
He used to light hunting fires to get kangaroo
Grass and wood, one man used to light for hunting
He used to make it go into the hunting area
He would be lighting fires
That one man
The mantis man is lighting fire
They were getting the animals for themselves
Lots of people went
There was one man in the middle
One man to the west, one to the north
One man to the south, they would be looking for kangaroo
They saw many kangaroo there
They go for all the kangaroo under the cover of smoke
They don't see those men for the smoke
Then the men would be hunting and spearing them with a spear
They speared and speared them, the kangaroos' bodies would be dropping and dropping
They took a stick from the river
They hit them on the neck
They picked up the carcasses
They carried the carcasses and put them here
They returned to camp with the carcasses
They returned with the animals to all the women
Then they would sit and relax
They speared the kangaroo for us
This is how they carried them, long ago
This man was me
I knew it ...

Part of a narrative of kangaroo hunting in Arnhem Land, told by Jackie Bun.ganiyal to Bowman *et al.* (2001).

8

Environmental change and Aboriginal history

THE ARRIVAL OF people caused two huge environmental changes in Australia: first the extinction of the megafauna, and then (very probably) a rearrangement of the vegetation in consequence of that extinction. Australian environments were transformed within a few thousand years.

We now need to consider what happened over the ensuing forty thousand years or so, up to the time that Europeans arrived. There is evidence that towards the end of this long interval – within the last six thousand years – Australian environments were reshaped by people for a second time. This time human use of fire was part of the cause, and it was associated with increased population size and a set of technological and cultural changes that resulted in an intensification of human impact on the environment. This chapter puts the changes of the last six thousand years into the context of archaeological and climate trends over the whole of the prehistory of Australia.

THE GLACIAL MAXIMUM AND THE GREAT FLOOD

When people came to Australia the climate was mild and stable, but it changed for the worse after 35 kyr ago. First, the LGM brought desertification, and this was followed by global warming and the flooding of vast territories as the sea rose through 130 metres. These stresses had drastic effects on the human population.

As the LGM approached, temperatures, rainfall and atmospheric CO_2 all declined. The combined effect was a suppression of plant growth and a decline of woody vegetation, which in turn would have meant fewer resources for hunter-gatherers. Water also became more scarce, and this may well have made large areas uninhabitable. The problem of lack of water was especially severe in the north of the continent and in the arid zone. In the tropical north the failure of the monsoon for perhaps 20 kyr around the LGM, together with the fact that temperatures and therefore evaporation remained relatively high, transformed a seasonally wet–dry tropical climate into a continuously arid one. Much of central Australia was only semi-arid when people first reached it, but deep aridity – waterless deserts, shifting sand dunes and very sparse vegetation cover – took hold in the centre of the continent and then spread outwards in a drying that began about 35 kyr ago and did not go into reverse until around 15 kyr ago.

The archaeology of northern Australia records a distinct decline in the number of sites occupied by people during the LGM. In the Kimberley region of the northwest, which probably became severely arid at the LGM, this effect was dramatic. The Kimberley has several archaeological sites with occupation histories reaching well back into the Pleistocene, but all were unused for thousands of years some time between 25 and 10 kyr ago (Walsh & Morwood 1999). There are sites elsewhere in northern Australia that were continuously occupied through the LGM, but the densities of artefacts in these sites typically dropped low in the driest period at the peak of the LGM and for several thousand years afterwards, presumably reflecting intermittent use if not complete abandonment (Morwood & Hobbs 1995; O'Connor *et al.* 1993).

The implication is that there must have been a general population decline in northern Australia, reaching catastrophic proportions in places like the Kimberley. A somewhat different interpretation is that there was a large-scale shift of population into more favourable habitats below present-day sea level. The climate conditions that brought thousands of years of drought to northern Australia also drove down sea levels and created new habitat on the Carpentarian Plain, much of it relatively well watered. Perhaps there was a northern Australian diaspora that settled on these low countries, to return to their homelands ahead of rising sea levels thousands of years later.

There appears also to have been a reduction in occupation in the arid zone. Sites that show signs of use through the LGM have low densities of artefacts and may have been abandoned for quite long periods. This could be true of two central Australian sites that were first occupied more than 30 kyr ago, Kulpi Mara and Puritjarra (Thorley 1998). There appears to have been less effect of the LGM on occupation through much of the semi-arid, subtropical and temperate regions of mainland southern Australia. The density of artefacts in archaeological sites in these regions tended to be low for the whole of the Pleistocene and remained so through the LGM (Lourandos 1997).

The pattern in Tasmania is different again. Many Pleistocene sites were continuously occupied through the LGM, only to be abandoned 12 or 10 kyr ago when much of the rest of Australia was emerging from aridity. The reason is that Pleistocene-age sites in Tasmania are concentrated in the southwest of the island, which under the cool and dry conditions of the LGM was covered by low shrublands and grasslands with scattered woodland patches (see Chapter 5). Despite the cold this was evidently a productive environment for hunters, and the Tasmanians appear to have specialised in the hunting of large mammals – especially the red-necked wallaby *Macropus rufogriseus* – to a greater extent than other Australians (Cosgrove & Allen 2001). With increasing rainfall, temperature and CO_2 in the latest Pleistocene, the open steppes of the southwest were overtaken by very dense cool temperate rainforest. This vegetation was largely impenetrable and was poor habitat for red-necked wallabies (Cosgrove 1995).

The ending of the LGM is marked by the onset of sea-level rises just after 20 kyr ago. The sea continued to rise until about 7 kyr ago, when it peaked at about one metre above its present stand. Rising seas drowned the connections between New Guinea, mainland Australia and Tasmania, and flooded close to three million square kilometres of land around the margins of the continent. In places where the sea rose over gently sloping continental shelves the rate of inundation of coastal land was very high. For example, on the northern Australian coast, at Princess Charlotte Bay in north Queensland, and on the Nullarbor coast, the coastline must have advanced at speeds of several hundred metres per year (Flood 1999; Smith 2005). This rate of loss of coastal territory was fast enough to have been a seriously disruptive factor within human lifetimes. Indeed, Cane (2001) suggests that some Aboriginal myths describing the resistance of spirit beings to encroachments of the sea may be an echo of these disruptions, just as the Genesis story of the flood might ultimately derive from direct experience of late-glacial rises in sea level in the Middle East.

The rising sea must have forced a constant relocation of settlements as people kept ahead of the advancing coastline, causing local crowding and the mixing of cultures at the front of the advance. There is evidence of such changes on the coasts of southern Australia (Cane 2001). At Allens Cave on the Nullarbor Plain, for example, there was an increase in the density of artefacts and appearance of new materials, including but not restricted to materials of marine origin, as the coastline drew in from 130 kilometres away to its present location 4 kilometres from the cave. After stabilisation of the coast, the density of artefacts at the site dropped to previous levels and its occupants settled back into a terrestrial economy.

PLATE 25 Toolache wallaby *Macropus greyi*, from John Gould's *Mammals of Australia* (1863).

PLATE 26 Pig-footed bandicoot *Chaeropus ecaudatus*, from John Gould's *Mammals of Australia* (1863).

PLATE 27 The desert bandicoot *Perameles eremiana* (middle) and its close relatives the western barred bandicoot *P. bougainville* (top) and eastern barred bandicoot *P. gunnii*. The desert bandicoot is extinct, the western barred bandicoot is extinct on the mainland, and the eastern barred bandicoot is almost so.
(painting by Frank Knight, reproduced courtesy of the artist and Oxford University Press)

This history suggests that rising seas caused a wave of population movement and a transient diversification of cultural activity.

Such population movements might also have triggered conflict. Tacon & Brockwell (1995) interpret changes in the density of artefacts in some Kakadu sites as reflecting increased population density as coastal people were driven into the hinterland by the advancing sea. And at around the same time they find depictions of fighting in the region's rock art, with people 'flailing boomerangs, dodging spears and chasing each other with weapons raised'. Frightening animal-headed beings also appear, and at some sites there are paintings of people arrayed in 'great battle scenes'. These are among the earliest depictions of warfare in any world art tradition (Tacon & Chippendale 1994).

The rising of the sea was caused by a warming climate that eventually brought higher rainfall, but not for several thousand years (Dodson & Mooney 2002; Kershaw 1995). Parts of southern Victoria still had desert vegetation, complete with red kangaroos, as recently as 12 kyr ago. This interval in which temperatures rose over a land that remained arid might well have been the most harsh period of the whole of the glacial cycle. The final breaking of the LGM drought happened at rather different times throughout the continent, but in general rainfall had risen to present-day levels by 10 kyr ago. Moisture and temperature continued to increase. They reached peak values at about 8 kyr ago in most of southern Australia, and perhaps 2 kyr later in the inland and north. These wet and warm conditions lasted until 4–5 kyr ago, in an episode known as the 'climate optimum'. In southeastern Australia mean temperatures during the climate optimum were probably about 1 degree C warmer, and rainfall was about 30 per cent higher, than now (Kershaw 1995). ENSO cycling was weak, so the climate was also unusually stable (Gagan & Chappell 2000; Tudhope *et al.* 2001), and atmospheric CO_2 levels had returned to their pre-glacial levels (see Figure 5.1).

Under these benign conditions there was more rainforest in the southeast and northeast, more complete vegetation cover and less sand dune activity in the dry inland, and greater tree cover generally, than at any time since. Also, rising sea levels created productive new habitats around the coast: shallow coastal seas, tidal flats, tidal reefs, estuaries and mangrove forests. In some places, such as the Alligator Rivers region of the Northern Territory, sedimentation behind the intertidal zone built levees which protected some coastal flats from saltwater inundation and led to the formation of freshwater wetlands and lagoons (Jones 1985). These complexes of near-coastal habitats reached their maximum extent about 6 kyr ago, although in the Alligator Rivers their evolution continued until about 1–2 kyr ago, and they provided new and productive habitats for people. The response to this is seen in a general proliferation of coastal middens dating to about 8–6 kyr ago (Cane 2001; Flood 1999).

POPULATION GROWTH

Figure 8.1 shows changes in the number of rock-shelters known to have been occupied by people from 40 kyr ago to the recent past. This compilation suggests that the density of occupation was generally low until about 10 kyr ago, when site density began to increase. Increasing site density has been interpreted by many archaeologists as indicating population growth (see Flood 1999; Hughes & Lampert 1982; Lourandos 1983, 1997; Lourandos & David 2002; Ross 1985).

This reading of the archaeological evidence could be confounded by two problems. First, the increase might reflect not an increase in number of people but a redistribution of the same population over a large number of sites, perhaps because of a reduction in group size or a rearrangement of movement patterns. This seems not to have been so, as there is also a very general increase in the density of artefacts within sites during the period that site numbers were rising in the Holocene – that is, both the number of sites and the intensity of use of individual sites seem to have risen (Lourandos 1997). Second, the apparent increase in site numbers through time might be a function of the disappearance of traces of earlier occupation. A constant rate of disappearance of evidence through time

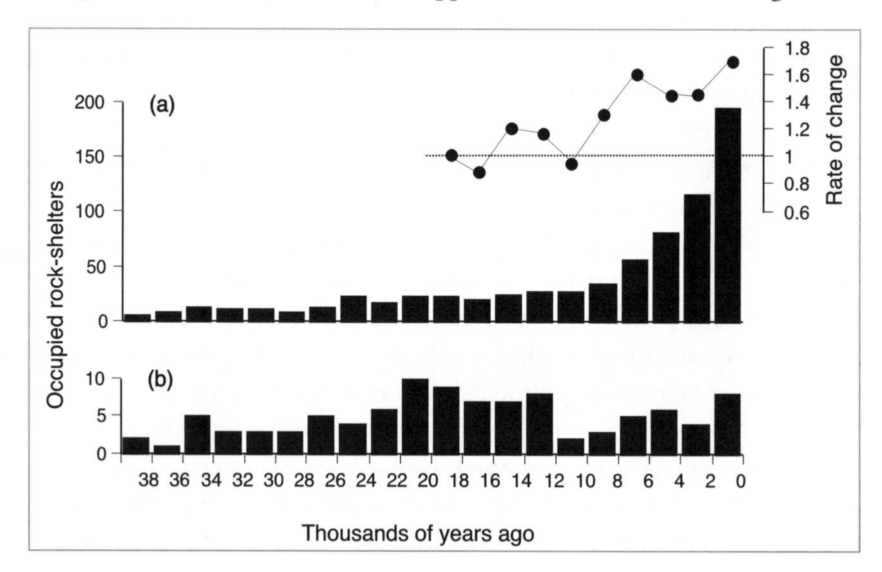

FIGURE 8.1. The number of rock shelters known to have been occupied in 2 kyr intervals from 40 kyr ago to the recent, in **(a)** Australia including Tasmania and **(b)** Tasmania only. The rate of change (●) in number of occupied shelters between successive intervals for the last 20 kyr is also shown for the mainland.
Notes: Rate of change was calculated by dividing the number of shelters occupied in one time interval by the number occupied in the preceding interval; the dotted line shows the value for an unchanging number of shelters. A shelter was classed as occupied in a given interval if an age determination on archaeological material fell within that interval. These figures were compiled from a database of archaeological dates created by me, and available on request.

would produce an exponential rise in the number of sites known in the approach to the present, quite like that shown in Figure 8.1. However, there are reasons to think that this sampling bias does not account for the recent increase. The increase is apparent on old land surfaces, showing that it is not produced solely by the wiping away of old sites by erosion or their burial under sediments (Ross 1985); also, the trend is very strong for rock-shelters, which presumably contain the most durable evidence of occupation, as well as for middens and open sites that are more readily disturbed and degraded (Lourandos & David 2002).

The first part of the Holocene increase, from 10 to about 6 kyr ago, could be explained as a consequence of two environmental factors: rising sea levels, which reduced the area of the continent and forced a concentration of settlement while creating productive coastal habitats, and an improving climate over the whole continent. Sea levels had stabilised by 6 kyr ago, and after 5 kyr ago there was a partial return to a drier and slightly cooler climate that put the finishing touches on the development of arid environments in Australia. These declines in average temperature and rainfall during the late Holocene were small in comparison to preceding climate changes (Dodson & Mooney 2002; Kershaw 1995), but perhaps the dominant influence through this period was a sharp increase in climate fluctuations due to a strengthening of ENSO cycling, to levels greater than any recorded for the previous 150 kyr (Tudhope *et al.* 2001). Continuation of the relationship between environmental conditions and the human population that seems to have applied before 6 kyr ago should have seen a late Holocene fall in population size, as people were pressed by drier conditions and buffeted by the increased variability of climate. But, it seems that something fundamental had changed, and population growth continued apace even as the climate became more harsh.

A HOLOCENE REVOLUTION: INTENSIFICATION

In fact, quite a few things had changed. Most obvious in the archaeological record is a diversification of stone tools. Until about 5 kyr ago stone artefacts in Australia consisted of a small range of predominantly general-purpose tools, with a working edge produced by striking a flake from a stone core, and without further refinement (Lourandos 1997; Mulvaney & Kamminga 1999). Most were probably used in the production and maintenance of wooden implements, and they had few primary or specialised functions. There was little change in toolmaking practices through the

Pleistocene, and there are remarkably few signs of technological responses to the challenges posed by the LGM.

But the mid-Holocene saw the appearance of many new tool types and a distinctive new culture of toolmaking that is sometimes known as the Australian Small Tool Tradition. These new tools were generally smaller and more specialised than the earlier ones, and they were typically made from raw materials of higher quality, which must often have been acquired by trade (Lourandos 1997). They include stone points, and 'backed flakes' (or 'backed blades') – small sharp-edged fragments with one blunted side forming the 'back' – of many distinct kinds. As well, a variety of new adzes, hatchets and many other tools appeared (Mulvaney & Kamminga 1999). Many of the new stone artefacts appear to have been components of composite tools. Backed flakes were probably mounted in rows on spears to serve as barbs, stone points were often used as spear-heads, and the other types were hafted. Backed flakes were the most numerous of the new artefacts. They became widespread around 4 kyr ago, but there is evidence that they originated in northern Australia as early as 15 kyr ago (Slack *et al.* 2004). Although the new approaches to toolmaking were geographically widespread, there were strong regional differences in the particular types and designs of tools that appeared. This suggests that the technological change was not simply the widespread adoption of a single new toolkit, but a flowering of technical innovation that produced many distinct local inventions in toolmaking.

At about the same time people began to use a wider range of animal and plant species. Use of these new resources often depended on new material culture, cooperative behaviour, and specialised knowledge. Morwood (1987) noted an increase about 4 kyr ago in the range of mammal species that were hunted in southeastern Queensland. Before 4 kyr ago, medium-sized macropods (the black-striped wallaby *Macropus dorsalis* and pademelons *Thylogale* sp.) were the main mammals hunted, but afterwards the range widened to include smaller-bodied species. Morwood suggested that in the earlier period mammals were killed mainly by lone hunters who ambushed, tracked or stalked animals and speared them, while in the later period hunters also worked in groups to drive smaller animals from dense undergrowth into hunting nets. In central and northern Queensland the use of toxic plants such as cycads, which can be eaten only after extended treatment, began around 3 kyr ago (Cosgrove 1996; David & Lourandos 1997). The grinding of grass seed to make flour may be even more recent, at least as a widespread activity (David 2002).

There are also signs of more sedentary and permanent occupation of habitats that had previously been used intermittently or seasonally. In southwestern Victoria networks of channels were constructed to join lakes and swamps for the purpose of trapping fish and eels. Mound systems were built to provide permanent dry ground that allowed the continuous occupation of swampy habitats (Lourandos 1983). The investment of time and effort that was needed to build and maintain these systems suggests that they were at the centre of essentially permanent settlements. Very large and relatively recent middens on Cape York also indicate intensive and continuous occupation of particular coastal sites (David & Lourandos 1998).

Art styles changed in ways that suggest Aboriginal societies became more distinct and strongly differentiated from one another during the last 5 kyr (Morwood 2002). Characteristic artistic traditions began to develop in different regions and, more recently, marked differences in the styles and subject matter of artworks arose over quite small distances within regions. For example, David (2002) describes the development of local art styles within the last 2–3 kyr around the Mitchell River of southeastern Cape York Peninsula: linear and geometric motifs predominate to the south of the river, and figurative painting to the north. Each area in the north has a distinct figurative style, and the boundaries between them are very sharp. On the Koolburra Plateau, for example, echidna–human shapes account for 28 per cent of paintings, but are never found in neighbouring areas. This spatial structure in art style could represent, at the larger scale, the ambits of formal trade networks that maintained parallels in the aesthetic and sacred traditions of the interacting groups, and at a finer scale the cultural distinctiveness of smaller local groups with their own special beliefs.

While local differentiation of social groups seems to have become stronger, there is evidence that different groups began to participate in large formal gatherings. These were often associated with trips to take advantage of seasonal concentrations of food. An example is the exploitation of summer aggregations of bogong moths *Agrotis infusa* in the highlands of southeastern Australia. The annual moth feasts in this region were the occasion for meetings and ceremonies involving hundreds of people from groups who at other times operated in separate territories, and they were an important forum for negotiating intertribal business (Flood 1999). Annual harvests of bunya pine nuts served similar social functions in southeastern Queensland (Morwood 1987).

How ancient these food festivals may be is difficult to say, but Lourandos (1997) suggests that large ceremonial gatherings of this kind may have begun only within the last few thousand years. David (2002) argues that throughout much of inland Australia the development of seed-grinding to make flour was a historical prerequisite for large gatherings.

Damper was a staple food for people at these meetings and large supplies of flour had to be laid by before an important ceremony could be held. This implies that seed-grinding stones might be an archaeological marker for the regular occurrence of large gatherings, and most grindstones are less than 2 kyr old.

Probably as a combined result of technological and cultural changes, the range of environments used by people increased as intensive settlement extended to regions that were previously marginal and had been used intermittently, if at all. The most important of these were the sandy deserts of central and western Australia (Veth 1989), mallee woodlands in southern Australia (Ross 1981), alpine and subalpine habitats (Flood *et al.* 1987) and rainforests (Cosgrove 1996; David & Lourandos 1997). These habitats are of low productivity for people and permanent occupation of them presents special challenges, which were met by new strategies of resource use such as the development of knowledge and technology for utilising seed resources in arid lands (Edwards & O'Connell 1995; Veth 1989).

The complex of changes described above is usually referred to as 'intensification', because it appears to have involved both more intensive use of resources and more intensive interactions among people. It has been studied by many archaeologists, and reviewed at book length by Lourandos (1997). Lourandos sees the various strands of intensification – new technologies, use of new resources, more intensive and permanent settlement, ceremonial gatherings of many people, greater closure and differentiation of societies along with development of formal trade networks, occupation of marginal environments, and population growth – as being closely related to one another, and driven by an underlying set of social factors. He argues that before the mid-Holocene social groups were small and wide-ranging, with low levels of exclusivity of membership and territoriality. This form of society evolved into one with more tightly organised and closed social groupings that remained within smaller and more rigidly defined territories. People in these more closed societies developed formal systems of alliance and exchange with their neighbouring groups, which were regulated through planned ceremonial gatherings that brought together large numbers of people at defined places.

Intensification of resource use was thus due to three factors. The development of closed societies made it possible for people to intensify their use of resources within defined boundaries; the formation of extended trade networks facilitated the transfer of materials and innovations among groups; and the problem of catering for large ceremonial gatherings made more intensive resource-use a necessity. The general result was the growth of more elaborate, technically sophisticated and productive strategies of hunting, gathering and resource management. Lourandos suggests that the process may initially have been triggered by improving environmental conditions early in the climate optimum, which led to closer settlement and more intensive interaction among people. But from then on the new cultural forces that had been set to work drove further social, economic and technological change independent of environmental control.

Lourandos's interpretation of social change in Holocene Australia has its critics, but whatever their stance in the 'intensification debate' Australian prehistorians agree that there were many significant changes in Aboriginal society in the late Holocene, and that an underlying feature of them was a stepping up of the influence of people on their environment. It may be that it was only during this period that purposeful management of the landscape by people began to have significant environmental effects. As a result of this more intense management, people were able to increase the productivity of their environment, and they began to leave an imprint on the landscape that, in some places, must have been more like that of farmers than of hunter-gatherers. The canal systems that were built in western Victoria to increase the harvest of eels are one instance of this.

Another is the evidence that intensive use of some food plants maintained them at near-agricultural densities. For example, people in southern and western Victoria made heavy use of the radish-like roots of the murnong, or daisy-yam *Microseris scapigera* (Gott 1983). As they dug up its roots, women and children loosened the soil and broke up clumps of the plant in a process that one European observer described as 'accidental gardening'; it is possible that the tops of the plant or a few roots were put back into the loosened soil where they would have regenerated. People also burnt murnong fields in the dry season, and this favoured the plant by suppressing competition. As a result, murnong grew very densely in its favoured sites. On one occasion George Augustus Robinson observed murnong-gatherers 'spread over the plain as far as I could see them', each woman with as heavy a load of roots as she could carry (Gott 1983). Murnong declined very soon after European settlement and it is now rare, mainly because sheep quickly learned to root it out of the loose soil in which it grew and lived on it until they had eaten it to local extinction. Also – as a consequence of the presence of sheep – Aborigines were excluded from murnong fields. The decline of the plant may have been hastened by the loss of its 'accidental gardeners'.

The one major exception to this picture is the Tasmanian Aborigines, who were cut off from the profound changes that swept through the mainland populations. The Australian Small Tool Tradition never reached the island, and the toolkit of the Tasmanians was much simpler than that of their mainland counterparts. They gave up fishing and stopped using bone points in the Holocene, so in some respects their technology and resource use actually became simpler (Jones 1995), and there is little or no evidence for the development of strategies to increase environmental productivity. The Tasmanian population seems not to have increased through the last 10 kyr in the spectacular way that the mainland population did (Figure 8.1b). Aboriginal population densities were evidently about five times lower in northern Tasmania when Europeans arrived than in comparable environments of southern Victoria (Lourandos 1997).

'FIRE IS FOR KANGAROOS'

The record of fossil charcoal suggests that fire increased in Australia through most of the Tertiary, and increased further during the last few glacial cycles. The rise in the importance of fire was driven mainly by climate – the long-term aridification of the continent and, perhaps, a relatively recent onset of climate instability due to ENSO. As shown in Chapter 5 there is very little evidence for an increase in burning as a direct and immediate consequence of human arrival in the middle of the last glacial cycle. Nevertheless by the time Europeans came to Australia, firing of the landscape was almost entirely at the hands of Aboriginal people. So, at some time in prehistory, people took over from natural sources of ignition as the cause of burning in Australia. There are two important questions about this takeover. What difference did it make to the pattern and ecological effects of burning? And, when did it happen?

These questions are difficult to answer. The most comprehensive recent review of the history of fire in Australia (Kershaw *et al.* 2002) concluded that, rather than imposing a fundamentally different fire regime on Australian environments, the effect of people was to accentuate a pre-existing climate-driven trend towards more fire and a more flammable and fire-adapted vegetation. But if Aboriginal burning simply pushed Australian environments in a direction in which they were already moving, it is very difficult to say how much of the recent development of fire-prone landscapes was due to people, and how much would have happened anyway. One could argue that Aboriginal people made no difference at all to the re-creation of Australia as the most flammable continent.

Horton (1982) made just this argument in response to Jones's (1969) concept of 'fire-stick farming' and Singh *et al.*'s (1981) conclusion that late Pleistocene increases in charcoal were directly caused by the arrival of people. Horton pointed out (uncontroversially) that in pre-human

Australia, fire regimes were determined by an interaction of climate, which ultimately determines the flammability of the vegetation in a region, and the rate of natural ignition of fires (mainly by lightning). Geographical variation in climate and rates of natural ignition produced variation across the continent in fire regimes and, therefore, in the degree to which vegetation was adapted to fire.

When people came to Australia they simply fitted in to this natural variation. They used fire in vegetation types that were prone to fire, and in these vegetation types human fire-lighting substituted for natural ignition sources without radically changing fire regimes or vegetation. In vegetation that was not prone to fire, people made less use of it. In neither case did they alter the pre-existing state of things, and 'had Aborigines never reached Australia, the distribution and adaptations of the plants and animals of the country would have been almost identical to those which the first European explorers found around two hundred years ago' (Horton 1982).

Horton's paper was written at a time when very little was known about the ecological effects of Aboriginal landscape burning. There is still surprisingly little known about this important topic, but Bowman (1998) provides solid evidence that traditional Aboriginal burning did, and in some places still does, have strong effects on the structure of vegetation and the abundance of particular plant and animal species.

The region where this is best studied is central Arnhem Land (Bowman *et al.* 2001; Bowman & Prior 2004; Bowman *et al.* 2004; Yibarbuk *et al.* 2001). Here, fires set by people shape the vegetation in very many ways. Most generally, burning maintains a generally open structure in widespread plant communities such as *Eucalyptus tetrodonta* woodland. Otherwise, traditional fire management using low-intensity fires produces a fine-scale mosaic of burnt and unburnt patches that maintains a high regional diversity of ecological communities, while at the same time protecting pockets of fire-sensitive vegetation such as *Callitris* woodland, sandstone heaths and rainforests. Planned fires reduce fuel loads in places such as creek lines that would otherwise act as corridors for the spread of destructive fires. Fire is carefully controlled to avoid burning trees in flower, so as not to interfere with the production of honey and fruit. After yam beds are dug, and a section of the top of each tuber is replanted in the soil, the spot is burnt to improve conditions for the plants when they re-sprout. Burning is also used to make resources more accessible, such as by clearing access to fishing holes and removing undergrowth around fruiting trees to make fallen fruit easier to collect. All of these practices depend for their effectiveness on the authority of knowledgeable people who understand the rationale for burning in particular areas and at special times, and on negotiation among people from neighbouring groups to ensure that their burning plans do not conflict.

All this sounds very much like firestick farming as Jones (1969) conceived it: the systematic and purposeful use of fire to reshape landscapes in ways that made them more productive and convenient for people. There seems no doubt that Aboriginal use of fire changed the ecological patterning of plant communities within intensively managed areas like central Arnhem Land, and it may well have altered the larger-scale distribution of vegetation types. An example given by Jones (1969, 1994) is the west coast of Tasmania, where large areas of open sedgeland that were maintained under Aboriginal burning quickly reverted to impenetrable *Leptospermum/Banksia* thicket when that burning ceased. Although there is still debate over just how much difference firestick farming made to Australian vegetation (see Benson & Redpath 1997), there is not much doubt that Aboriginal burning was an important force for ecological change.

But when did it first become important? As I argued in Chapter 5, not when people first arrived, and probably not for many thousands of years thereafter, during which time the vegetation may actually have become less flammable as a consequence of extinction of the megafauna (Chapter 7). Several long-term charcoal records agree in showing an increase in fire around the LGM, but there is nothing remarkable about this because an increase in fire might have been expected under the dry conditions of the time, and there had been similar increases in the absence of people at the previous glacial maximum.

The Holocene, on the other hand, does stand out as having unexpectedly high charcoal, whereas charcoal was mostly low during the previous interglacial. Kershaw *et al.* (2002) reviewed charcoal in Holocene records from 58 sites in southeastern Australia, and showed that charcoal levels tended to increase throughout the Holocene. Charcoal peaks from individual records were concentrated in two periods: between 6 and 4 kyr ago, and after 2 kyr ago. The first of these periods of charcoal increase began during the climate optimum, when rainfall was relatively high and ENSO variability was low. The effect of that climate should have been to reduce the prevalence of fire, not to raise it.

So, the tempo of burning in Holocene Australia does not seem to have been closely related to climate. What it does correlate with is human population growth. As Figure 8.2 shows, the periods with many local charcoal peaks coincided with times when the abundance of archaeological evidence of human occupation was increasing most quickly. Increased burning at those times might have been a simple consequence of more

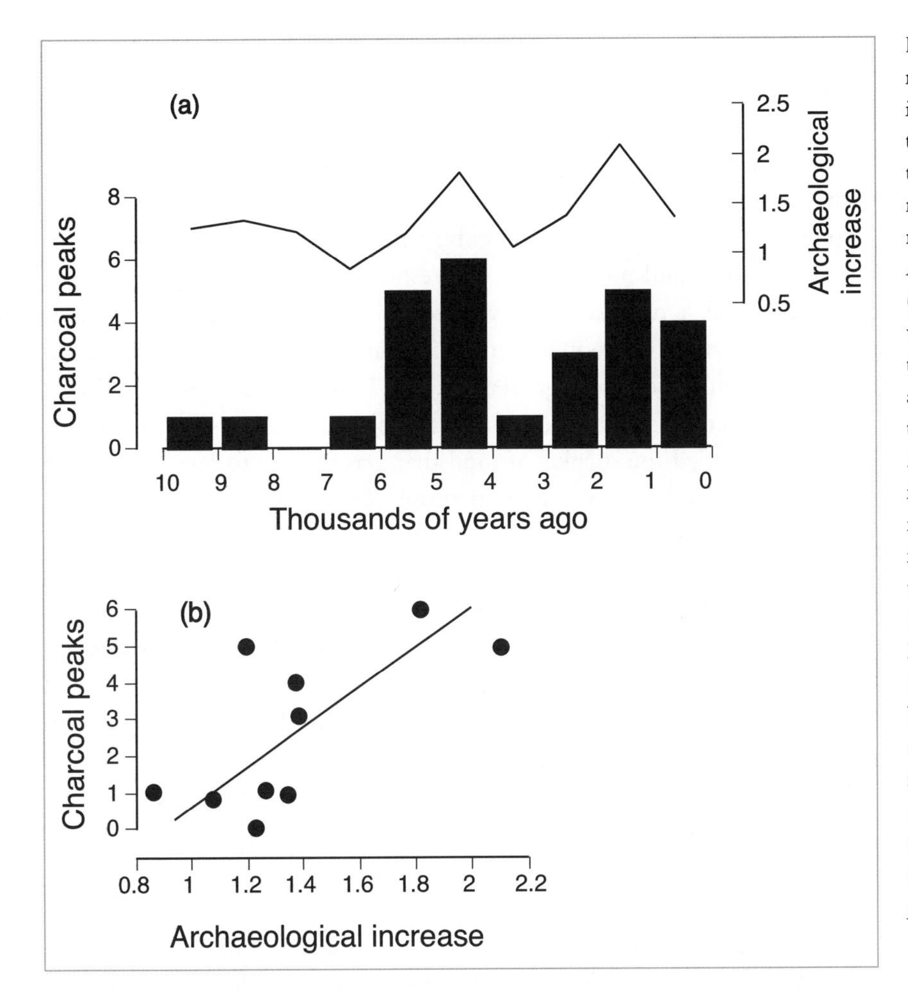

FIGURE 8.2 **(a)** The number of charcoal peaks in 1000-year intervals over the last 10 kyr (excluding the European period), as recorded in 58 charcoal records in southeastern Australia by Kershaw *et al.* (2002), in comparison with the rate of change in total number of dates on archaeological material through the same period. *Notes:* Rate of increase is measured as the ratio of number of dates in each interval to the number in the preceding interval (total number of Holocene dates = 1868). **(b)** The direct relationship between charcoal peaks and archaeological increase (the regression of charcoal peaks on archaeological increase is significant, $F_{1,8} = 7.02$, $p < .05$).

intensive occupation of the landscape by people, but it is also possible that there were important developments in the uses of fire that were related to the fundamental social aspects of intensification described above.

One of the most common goals of Aboriginal fire management was to increase the harvest of large mammals. Fire was used to maintain grassy habitats that were good for kangaroos, and fire was also used in the hunting of those kangaroos. Aboriginal fire management encompassed a huge range of purposes and strategies, but it seems that this particular role often took priority. This was made clear by the senior Arnhem Land men interviewed by Bowman *et al.* (2001): 'Fire is for kangaroos' (Mick Kubarkku); 'Just for kangaroo hunting. Fire. That's what it's for' (Djungkidj Ngindjalakku).

Large numbers of kangaroos could be killed in communal hunts that used fire. These were well-planned operations conducted in places that had been made attractive to kangaroos by earlier burning. Fires would be lit to drive the animals into concentration points, where hunters waited to spear or club them. The few firsthand accounts of these activities that have been recorded emphasise the care that went into their planning, and the importance of the rules that governed who had the authority to direct them (Bowman *et al.* 2001). These rules were crucial, as fire drives often involved cooperation among neighbouring clans, and failure to follow correct protocol and procedure could result in inter-clan conflict.

Communal kangaroo drives were often conducted along with ceremonial gatherings of people from different clans, and they formed an important element of those gatherings. The hunting activities were themselves closely connected to the sacred traditions affirmed in the ceremonies, and they provided the surplus of food that was needed to feed the people who had gathered to participate in those ceremonies. Lourandos (1997) identified the development of well-regulated cooperative behaviour and the holding of large ceremonial gatherings as two key elements in his explanatory model for intensification. These elements came together in kangaroo drives, and fire was the tool that made large-scale kangaroo drives possible. Perhaps this explains the connection between increased burning and archaeological evidence for change in the human population in Holocene Australia.

CONCLUSIONS

At the risk of oversimplification, the prehistory of Australia can be divided into two distinct ecological periods: before and after 6 kyr ago. In the early period the size of the human population appears to have been directly determined by environmental conditions. People lived where resources and conditions were favourable to them, and the human population changed its size and distribution in direct response to changes in the quality of the environment. In most respects hunting and gathering were relatively unspecialised and based on simple technology. People caused megafauna extinction in the early part of this period, but not because of any special effort or adaptation towards large mammal hunting. The megafauna went extinct because they were the component of the biota that was most sensitive to the rather small pressure that early Aborigines imposed on the environment of Australia. Other than that, mammal populations were probably little affected by the presence of people. Fire regimes in Australia varied through this period and fire become more prevalent, but this was primarily because of changes in climate.

In the period after 6 kyr ago the Aboriginal people of mainland

Australia began to exert more control over their environment. People increased environmental productivity by explicit and purposeful strategies of resource management that employed a wide range of new technologies. The result was that the human population became less dependent on simple climatic factors. Population size increased steadily through a time when the climate was becoming drier and more unstable, and a component of this increase was a re-invasion of the arid zone, which had been largely abandoned during the LGM. Fire was increasingly used in ways that reshaped the vegetation to make it more useful to people, and especially to increase harvests of large mammals. The impact of Aboriginal people on Australian ecosystems increased steadily through this whole period in a process that was still underway when Europeans arrived. The extent to which this intensifying impact was felt by mammal populations is explored in the next chapter, but that chapter must begin by describing another momentous event in the history of Australia's mammal fauna.

9

Dingoes, people, and other mammals

MIDWAY THROUGH the Holocene a new element was added to the cultural and environmental landscape of Australia. The dingo *Canis lupus dingo* has been regarded as a powerful introduced predator that caused the extinction of the mainland's two largest native carnivores, and as a cause of fundamental cultural change in Aboriginal Australia. I suspect that on both counts the dingo's influence has been overrated, and it has drawn attention away from other pressures on Australian mammals, the most important of which was 'intensification' (see Chapter 8). Nonetheless, in the transition from Aboriginal to European Australia the dingo emerged as perhaps the most ecologically significant mammal species on the continent. The changing role of the dingo in Australian ecology is a theme that will recur later in this book. For now, we need a careful evaluation of how the dingo fitted into Australian society and environment before the arrival of Europeans, and in particular how it was related to the decline of the thylacine and devil. This chapter begins by providing this appraisal of the dingo; I then turn to the larger question of the impacts of people on native mammals in Holocene Australia.

HISTORY OF THE DINGO

The dingoes of Australia are the descendants of early domestic dogs of Asia, which were descended in turn from Eurasian wolves (Savolainen *et al.* 2004). Eastern Asia appears to have been the single site of domestication of dogs, about 12 kyr ago (Leonard *et al.* 2002; Savolainen *et al.* 2002). Living dingoes represent an early stage in the process of domestication, before differentiation of breeds. The low level of genetic variation in the Australian dingo population suggests that it is derived from an isolated colonisation event, conceivably involving a single pregnant female, with no subsequent admixture from other populations (Savolainen *et al.* 2004).

When did dingoes come to Australia? The fossil record suggests just after 4 kyr ago. There are four reports of dingo remains aged between 3 and 4 kyr, at widely separated localities in southern Australia: 3.75 kyr at Madura Cave on the Nullarbor Plain (Milham & Thompson 1976); 3.44 kyr at Wombah Midden in eastern New South Wales (McBryde 1982); about 3.4 kyr at Fromms Landing on the Murray River in South Australia (Mulvaney *et al.* 1964); and about 3 kyr at Balmoral Beach, Sydney (Attenbrow 2002). Slightly younger remains occur at Ngurini Rock-shelter on the Mitchell Plateau in the western desert (2.9 kyr, Veth 1996) and Thylacine Hole on the Nullarbor Plain (2.2 kyr, Lowry & Merrilees 1969). Younger dingo remains are common in caves and archaeological sites throughout Australia. A slightly earlier date of arrival is implied by the range of genetic variation in living dingoes, which is consistent with isolation of the Australian population for about 5 kyr, although this could be an overestimate if there was significant variation within the small founding population.

A date of just after 4 kyr ago for the landing of Australia's ancestral dingoes places that event at the beginning of the expansion of Austronesian people – the ancestors of modern-day Pacific Islanders – from southeast Asia. These seafaring people eventually spread all the way to Easter Island in the east and New Zealand in the south (Hurles *et al.* 2003). They carried dingoes with them, possibly as a source of fresh food (Corbett 1995). Some of these travelling dogs might have wandered into the bush, or perhaps were given to Aborigines as gifts, if their owners made landfall in northern Australia. The transporting of dingoes to Australia is often taken as evidence that there was significant interaction between Asians and Australians 4 kyr ago, but the genetics of Australian dingoes suggest the contact was actually very limited. In fact, we need only imagine a lucky animal or two surviving the sinking of one boat near the coast of northern Australia to account for the genesis of Australia's dingo population.

Dingoes spread quickly throughout mainland Australia and they now occur in practically all habitats (Corbett 1995), but they never reached Tasmania. Perhaps the first dingoes established a wild population and spread through the wild, but it is possible that their expansion over mainland Australia was helped by people. These early dingoes were already in the first stages of domestication, so they may quickly have established themselves as companion animals to Aborigines. A model for just how quickly this could have happened comes from observations of the acquisition of modern breeds of dogs by Tasmanian Aborigines (Jones 1970). During the first few years after British settlement in Tasmania in 1803 the colonists were forced to hunt kangaroos and emus to survive, with the aid of dogs. The Tasmanian Aborigines, having had no experience whatever of dogs (or dingoes), were impressed by what they saw, and within a few years some 'wild' Aborigines had their own packs of dogs. By 1830 Aborigines over practically the whole of the island, including people who lived hundreds of kilometres from the colonies and had never seen a white man at close quarters, owned dogs. Dogs were owned by individuals who gave them names, and they were spread from one group to another by trade, as gifts and by theft; also, pups of feral dogs were captured and tamed. Dogs quickly become one of the most significant items of wealth for Tasmanian Aborigines and within thirty years they had entered the mythology of the people – who believed, for example, that after his death a man could continue to influence his dog's behaviour and personality (laziness in a dog was often explained in this way).

DINGOES AND PEOPLE

The traditional relationship between Aborigines and dingoes is poorly documented, mainly because dingoes were quickly replaced as camp dogs by European breeds. These were often valued more highly – according to Meggitt (1965) 'even the oddest mongrels' were preferred over dingoes. Descriptions by early Europeans of the nature of the association between Aborigines and dingoes, compiled by Meggitt (1965), are therefore brief and fragmentary. Aborigines hunted and ate wild dingoes, but the contribution of dingoes to their diet seems to have been variable and was probably not great overall. Hunting of dingoes more often consisted of taking pups from dens, the best of which would be raised in camp while any small or weak ones were eaten. In these expeditions the bitch would sometimes be found and killed but often was left to breed again. This therefore amounted to a system of harvest, and the practice was adapted for profit when colonial governments began paying bounties on dingo scalps (a pup's scalp was worth as much as an adult's). Some observers commented on the care given to pups, which even extended to suckling by women. Adult dogs were useful in camp to warn of the approach of strangers, and were valued for their company. They were sometimes used in hunting, where their role was to scent prey and to run down large game such as kangaroos and emus.

Views on the value of dingoes to the hunt differed strongly. Lumholtz (quoted by Meggitt 1965) wrote that in north Queensland the dingo was 'very useful to the natives, for it has a keen scent and traces every kind of game; it never barks, and hunts less wildly than our dogs, but very rapidly,

PLATE 28 Three desert bandicoots: (from the top) the greater bilby *Macrotis lagotis*, the lesser bilby *M. leucura* and the pig-footed bandicoot. The greater bilby is now rare, the other two extinct. (painting by Frank Knight, reproduced courtesy of the artist and Oxford University Press)

PLATE 29 Thylacine *Thylacinus cynocephalus*, from John Gould's *Mammals of Australia* (1863).

PLATE 30 White-footed tree-rat *Conilurus albipes*, from John Gould's *Mammals of Australia* (1863).

frequently capturing the game on the run'. By contrast, Major Thomas Mitchell, who travelled widely through the interior, commented of dingoes that 'we saw few natives who were not followed by some of these animals, although they did not appear to be of much use to them'. Meggitt's (1965) own experience of life with Warlpiri hunters in central Australia was that their dogs made little difference to hunting success. He suggested that dingoes had provided most aid to hunting in dense forest, and least in open habitats. Consistent with this, Jones (1970) assembled evidence that the domestic dogs acquired by Tasmanian Aborigines made the hunting of kangaroos and wombats much more effective, while his observations on Aboriginal hunters in more open country in Arnhem Land confirm that dogs were of little use to the hunt. The New Guinea singing dog, a near relative of Australia's dingoes that was brought to New Guinea perhaps 5 kyr ago, was also a valued hunting partner in dense forest (Bulmer 2001). Perhaps the most consistent contribution of dingoes to hunting was their ability to locate arboreal mammals, such as tree kangaroos and possums, by smell.

Some early observers commented that camp dingoes were well cared for, even to the point that excessive pampering explained their inadequacies as hunting dogs; others said that they were treated with a combination of sentimental affection and casual indifference to their welfare, and were maintained in a state of near starvation. It does seem plain that many dingoes that were taken as pups and raised as camp dogs eventually returned to the wild, although some were deliberately crippled to prevent this. Able-bodied camp dingoes were especially likely to abscond during the mating season. Breeding among camp dingoes was probably not significant, and they cannot be considered to have been truly domesticated. The morphological conservatism of Australian dingoes through the last 3.5 kyr, and their similarity to ancestral wolves, makes it clear that they underwent little or no selective breeding based on their utility to people. Rather, the status of dingoes in pre-European Australia was an odd mixture of the wild, the feral and the semi-domestic. Dingoes featured in many myths, songs and art works, and they had sacred status as totem animals. They also figured in many folk beliefs – the Adnyamathanha people of the Flinders Ranges, for example, believed that women and girls should not walk on the scratch marks of a dingo as this would cause soreness of the groin (Tunbridge 1991). There is no doubt that dingoes had deep cultural significance, regardless of their somewhat doubtful practical value.

The complex interaction of human and dingo populations in Aboriginal Australia makes it difficult to interpret the effect that each had on the other. There are two important questions. What effect might dingoes have had on human population size? And what effect might people have had on the size of the wild dingo population after the animal became widespread?

The answer to the first question is, probably, not much. The arrival of the dingo added a new item to the Aboriginal diet, but densities of wild dingoes were probably low (see below) and camp dingoes were generally not kept to be eaten, so the direct contribution of dingoes to the Aboriginal diet must have been small. Dingoes might have increased hunting success, and in some places they almost certainly did so; but as explained above there are also strong indications that for many, perhaps most, Aboriginal tribes dingoes added little value to the hunt. Perhaps interaction with dingoes stimulated cultural changes that led eventually to new systems of resource use and intensification. Many prehistorians have wondered about this, and Flannery (2004) recently put the idea in bold terms, describing intensification as a 'dingo-driven revolution'. I find this hard to accept, for several reasons. First, it is just as easy to argue that the spread of dingoes was a consequence of some cultural features of intensification – a new interest in novelty and variation, and extended trade networks – as to claim it as a cause of intensification. Second, dingoes arrived in the middle of the trajectory of Holocene population increase, not at its beginning. Third, the genesis of the Australian Small Tool Tradition precedes the arrival of the dingo (see Chapter 8). The date of about 4 kyr for the establishment of the dingo and the appearance of new tool types throughout the continent does suggest that there might have been a common factor in the rapid expansion of both – this factor might have been the development of formal trade networks.

The effect of people on the wild dingo population was probably to suppress it. This could have been by direct hunting, and, more importantly, by theft of pups, which would have reduced recruitment into wild populations. Most observers agreed that whether or not some dingoes were valuable as hunting dogs, people kept more dingoes than they needed, and that for the most part these dingoes were stolen from the wild as pups. The natural abundance of dingoes in Australia is low, and may have been considerably less than the density of the human population. In Kakadu National Park, for example, dingoes live at a density of about 0.14 per square kilometre (Corbett 1995), while the pre-European density of people in the same area was about 0.77 per square kilometre (Lourandos 1997). The combination of hunting and pup-stealing by a numerically dominant human population may well have been the most significant control over the demography of wild dingoes, and kept their populations small. On the other hand, the fact that camp dingoes filtered back into the wild would have tended to stabilise numbers in wild dingo populations, and may have provided a buffer which helped to maintain them in poor habitats and bad years. The demographic impact of people on wild dingoes would have increased with the growth of the Aboriginal population over the last 4 kyr.

People also competed with wild dingoes for food. Dingoes have broad and flexible diets, but they do prefer medium-sized and large mammals and they are expert hunters of kangaroos. In this activity wild dingoes were in competition with Aboriginal men. Sometimes this competition was direct. Meggitt (1965) recorded that one hunting method used by Warlpiri men was to track wild dingoes that were running down a kangaroo, allow them to make the kill, and then steal the carcass. Some offal would be left behind for the dingoes, with a view to transacting the same rather unequal business with them in the future. But whether or not dingoes and men encountered each other on the hunt, predation by Aborigines would have reduced the abundance of prey for dingoes. Krefft (1866) wrote of the Aborigines around the Murray-Darling junction that they 'hate the dingo most cordially for his living on the fat of the land, kill him on every opportunity and eat his flesh', a comment implying that they regarded dingoes as competitors. The effect on dingoes of competition for food from people would also have increased through time as Aboriginal hunting methods became more effective and their population grew. People may also have restricted the access of wild dingoes to water – one method of dingo hunting was to spear them at waterholes (Hamilton 1972) – and to rock-shelters. Corbett (1995) suggested that wild dingoes were rare when Europeans first came to Australia, noting for example that the explorer John McDouall Stuart rarely saw or heard them as he crossed central Australia three times between 1860 and 1862. This rarity could well have been a consequence of the various ways in which wild dingoes were subordinated to the human population.

MAINLAND EXTINCTION OF THE THYLACINE AND DEVIL

The thylacine and devil were both widespread in late Pleistocene Australia. The thylacine probably occupied the whole of the continent, and the devil occurred everywhere except the central and western deserts (Gould *et al.* 2002). Both species disappeared from the mainland during the Holocene, and the dingo is usually blamed for this. The case against the dingo rests on one piece of circumstantial evidence: both marsupial carnivores disappeared from the mainland after the arrival of the dingo, but they survived in Tasmania. Corbett (1995) suggests that dingoes out-competed thylacines, mainly because their ability to hunt cooperatively gave them an advantage in killing large prey like kangaroos. Thylacines, by contrast, are thought to have hunted alone, and in any case they are often portrayed as rather slow-moving, leaden and inflexible predators. The dingoes' capacity to operate in packs may also have allowed them to more effectively defend crucial resources, like carcasses and water. For these reasons, dingoes would have had the advantage over thylacines when resources were scarce. The same argument applies to the ecological interaction between dingoes and devils. Another possibility is that dingoes introduced a disease, possibly toxoplasmosis or something similar, which wiped out the other two species.

As most writers have accepted Corbett's (1995) views, especially the version that has dingoes out-competing thylacines and devils, it is worth thinking through the weaknesses of the model.

TIMING

Holocene dates on Australian thylacines are summarised in Table 9.1. There are seven that appear to be reliable; two of these are direct radiocarbon dates on dry tissue from mummified thylacines found in caves on the Nullarbor Plain (Thylacine Hole and Murra-el-elevyn Cave), the youngest of which is around 3.4 kyr old. Archer (1974) obtained a slightly younger age of about 3.3 kyr for thylacine remains in Murray Cave near Perth in the southwest of Western Australia, but the association of the dated material (charcoal) with the animal remains in this case was not very precise. There are other thylacine remains of probable Holocene age, but without precise dates. Wakefield (1967) found thylacine remains in what he considered to be Holocene deposits in McEachern Cave in southwestern Victoria, and a thylacine that was apparently 'not very old' in Lava Cave in the same region (Wakefield 1963); a thylacine was found in aggregated cave rubble of probable Holocene age in Monajee Cave from the Pilbara region of Western Australia (Kendrick & Porter 1973); and O'Connor (1999) recorded a thylacine in archaeological levels of probable mid-Holocene age from Widgingarri Rock-shelter on the Kimberley coast. At Madura Cave on the Nullarbor Plain thylacine remains were found in sediments overlying a level dated at 3.75 kyr and which contained dingo remains, but Milham & Thompson (1976) provided little detail on the thylacine material. The same site contains isolated remains of other extinct fauna

Age (kyr)	Site	Source
8.1	Horseshoe Cave, Nullarbor Plain	Archer (1974)
5.1	Thylacine Hole, Nullarbor Plain	Lowry & Merrilees (1969)
~4.9	Devon Downs, Lower Murray River, SA	Smith (1982)
~4.3	Fromms Landing, Lower Murray River, SA	Mulvaney *et al.* (1964)
4.1	Horseshoe Cave, Nullarbor Plain	Archer (1974)
3.5	Murra-el-elevyn Cave, Nullarbor Plain	Partridge (1967)
~3.4	Devon Downs, Lower Murray River, SA	Smith (1982)

TABLE 9.1 Holocene dates on thylacine remains from mainland Australia

judged on the basis of their chemical composition to have been redeposited from older sediments, so the reliability of this post-3.75-kyr date for thylacines is in doubt. Thylacine remains occur in human occupation levels of Holocene age in New Guinea, at Nombe Rock-shelter (Plane 1976) and at Kiowa Rock-shelter (Plane 1976; van Deusen 1963).

So, the firm dates for thylacines in Holocene Australia are concentrated in a small region of the south coast from the Nullarbor Plain to the Murray River (Figure 9.1). The evidence for a close correlation of dingo arrival with thylacine extinction is restricted to this area. In contrast, late Holocene occurrences of the dingo are widespread, as noted above. Late Pleistocene thylacine sites are also widespread: they include Cloggs Cave in eastern Victoria (Flood 1973), Lakes Menindee and Mungo in western New South Wales (Hope 1978; Tedford 1967), southwest Western Australia (Balme *et al.* 1978), the western Kimberley (O'Connor 1999) and Mandu Mandu Rock-shelter on the Pilbara coast of Western Australia (Morse 1993). Also, thylacines are depicted in rock paintings and some rock engravings over much of northern and northwestern Australia.

These depictions of thylacines in rock art provide an independent source of evidence on the time of their decline in the north. In Arnhem Land thylacines are a common subject in paintings from two ancient art styles, the 'large naturalistic animal' style and the subsequent 'dynamic figure' style (see Plate 17; Chaloupka 1993; Murray & Chaloupka 1984). They were especially significant in 'dynamic figure' paintings, where they represent just over 9 per cent of the animals depicted, ranking third in importance behind fish and macropods (Tacon & Chippindale 2002). The 'dynamic figure' art style is thought to be about 13 kyr old (Tacon & Brockwell 1995). It includes no definite depictions of dingoes, which do not appear in Arnhem rock art until the more recent 'X-ray' art style (Flood 1997), which was painted within the last 4 kyr.

FIGURE 9.1 Localities with thylacine occurrences of definite (●) and probable (○) Holocene age on mainland Australia (see text for sources).

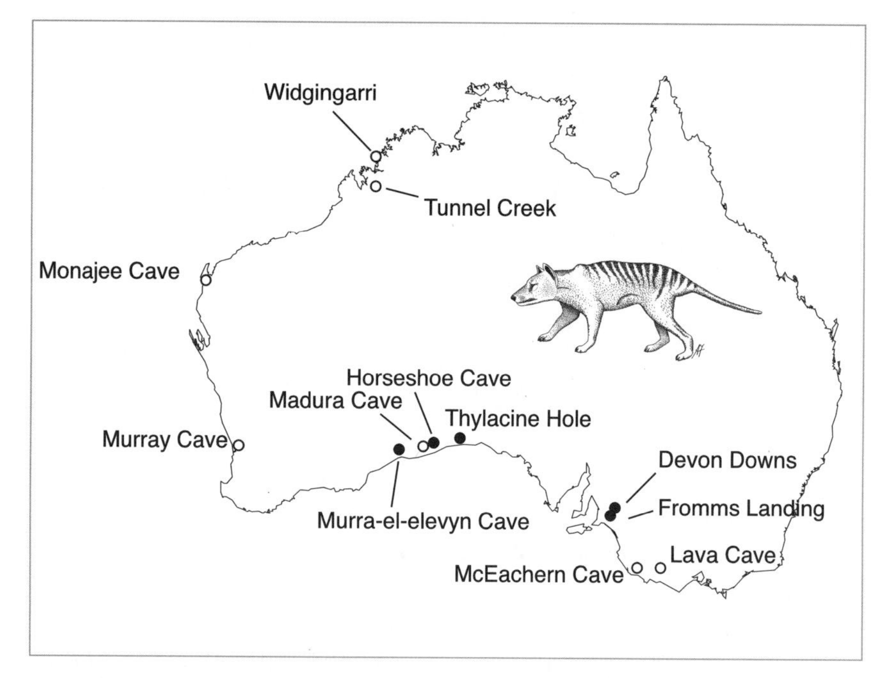

Thylacines also appear in early rock art styles of the Kimberley, specifically those referred to by Walsh (2000) as the 'irregular animal infill' period and the later 'clothes peg figure' period. These art styles in some ways parallel the naturalistic animal and dynamic styles of Arnhem Land, although they may be somewhat earlier in date, and both might pre-date the Last Glacial Maximum (Morwood 2002). Dingoes do not appear in them, and according to Walsh (2000) 'there is no depiction of any quadruped which could be considered to represent the dingo in Kimberley rock art until the very recent times of the late Wandjina period', which may well be confined to the last thousand years or so. The 'clothes peg figure' art is separated from the Wandjina paintings by the long 'clawed hand' period, which is thought to have extended through the early- and mid-Holocene. There are many animal depictions in this art style – echidnas are especially common – but no thylacines or dingoes (Walsh personal communication).

To summarise, while there is good fossil evidence that thylacines were extinct on the mainland soon after 3.5 kyr ago, shortly after the time when dingoes appear throughout Australia, this extinction date may apply only to part of the far south of the continent. We cannot be sure that the arrival of dingoes was linked to thylacine extinction elsewhere, and the rock art of northern Australia actually suggests that thylacines were gone before dingoes arrived. The fact that people stopped painting thylacines at a certain time does not prove their extinction, but it is quite striking that, even though the prominence given to thylacines and dingoes in paintings at different times shows that both species were significant to people, there are no cases where paintings of the two species are closely associated and rendered in the same style, and certainly none that illustrate any interaction between them. It is possible that the decline of thylacines was gradual and that it was completed in the north thousands of years before the arrival of dingoes.

There may have been a similar north-to-south decline in the devil. The youngest devil remains from the mainland have been dated to 430 years before present in the southwest of South Australia (Archer & Baynes 1972). The association of the devil specimen with the material dated by Archer and Baynes was tenuous, but they supported this recent date with reference to a nearby site at which devil remains were dated to 620 years before present (but, unfortunately, without providing details). At Lava Cave in western Victoria, Wakefield (1963) found remains of 60 individual devils, of which 'many specimens appeared to be quite modern. One mandible still had pieces of dried tissue adhering to the bone.' Devil bones of recent appearance have been found at several other sites in southern Victoria (Gill 1953). There have been historical records of living devils in mainland Australia (Paddle 2000). These are almost certainly not valid, but the evidence from Archer and Baynes (1972) and Wakefield (1963) of late survival makes it seem plausible that some devils were still alive in the most southern parts of mainland Australia when Europeans arrived. The youngest date for a devil in northern Australia comes from an archaeological site, Padypadiy on the East Alligator River, where devil remains occur just above the base of the deposit dated at 3.12 kyr (Calaby & White 1967). Devils appear along with thylacines in rock art from northern Australia, although less frequently, but they too are absent from the more recent art styles that contain dingoes.

THE DINGO AS SUPERIOR COMPETITOR?

The dingo and thylacine were both hunters of large mammals, but they differed in a number of ways. The thylacine was evidently a specialised killer of large prey. This is shown by its tooth and skull morphology (Wroe 2002; Wroe *et al.* 2005), and by the fact that Tasmanian thylacines evidently preferred to eat meat that they had freshly killed themselves (Guiler 1985). The dingo has a broad diet that includes many small animals as well as vegetable matter, and it is a ready scavenger. This suggests that there was scope for the two species to have coexisted by using different food types. A large population of dingoes living alongside thylacines would presumably have reduced the abundance of large prey, such as kangaroos, and this situation would have been more dangerous for the specialised thylacine than for the generalist dingo, which could have compensated by switching to alternative prey or feeding on carrion. However, as I argued above there are reasons to think that people held wild dingoes at low numbers, so the ecological effects of dingoes on thylacines may well have been small.

There are three other problems with the idea that thylacines lost out in competition with dingoes. First, the contrast in their hunting behaviour may not have been as great as is often supposed. Guiler's (1985) review concluded that thylacines hunted alone and made kills either by ambush or dogged solitary pursuit. This has been disputed by Paddle (2000), and the difference in views is interesting. Guiler's opinion was based largely on twentieth-century accounts which, Paddle argued, recorded the behaviour of animals that had already been dispersed, reduced in number and stressed by heavy persecution. Instead, he sought to uncover all nineteenth-century observations of thylacine behaviour, and his piecing together of this evidence revealed a flexible and powerful predator that hunted large prey in quite similar ways to dingoes. One observer even wrote that thylacines hunted in groups 'with the pertinacity of a pack of wolves on the steppes of frozen Russia'.

Thylacine groups consisted of an adult pair and their young, which remained with them until they were close to adult size, evidently departing before the subsequent litter left the pouch. While considerably smaller than the largest dingo packs, these thylacine groups hunted in a coordinated fashion. A hunting pair would work together, one animal setting the prey moving and the other making a final dash to bring it down. Older pups would also work with their parents, helping to run prey towards another group member waiting in ambush. Mr G. Stevenson, whose family had been successful thylacine trappers, wrote that 'The natural food supply of the tiger is kangaroo & wallaby, chiefly in the tea-tree scrubs, there 2 or 3 will get together, one will crouch down beside a track, and the others will hunt the wallaby, and when one comes along the tiger will pounce on it' (Paddle 2000). Early accounts portray thylacines on the hunt as being fast and nimble.

Second, thylacines were big. At around 30 kilograms a thylacine was about twice as heavy as the average dingo, and this should have given it the upper hand in any fair fight. Paddle (2000) and Guiler (1985) also comment that in Tasmania dogs often showed fear of thylacines. Thylacines bailed up by hunting dogs would fight hard in self-defence. For example, the *Hobart Town Gazette* of 6 December 1817 recorded the killing of a thylacine suspected of sheep-killing, which was 'attacked by seven dogs, and made a stout resistance, till at length it was killed with an axe by the stock keeper' (Paddle 2000).

Finally, there is the fact that thylacines persisted for over a hundred years in Tasmania in the presence of wild dogs. Dogs became feral in Tasmania soon after they were brought to the island by British settlers. The feral dog population grew quickly, wild dogs killed more sheep than did thylacines, and the wild dog population was probably much larger than the thylacine population. But there is no evidence that thylacines declined as wild dogs increased. If anything, the reverse happened: thylacines were rare when the British first settled Tasmania, but may have increased later in the nineteenth century. The catastrophic decline of thylacines in the early part of the twentieth century was not related to wild dogs (see Chapter 11).

It is a little easier to imagine how dingoes could have put pressure on devils. As scavengers, dingoes and devils might have been direct competitors. Also, recent experience in Tasmania shows that dogs can easily kill devils and may even have a propensity to do so (Jones *et al.* 2003). Devils defend themselves against dogs by backing against a tree trunk or some other solid object and making static threats. This may deter a frontal attack by a single dog, but a pair of dogs can attack from the front and sides (Menna Jones personal communication). Lacking cooperative behaviour, devils are easily overpowered and killed by dogs attacking in this way. Probably, groups of dingoes could easily have dominated devils and excluded them from carrion.

A ROLE FOR PEOPLE?

Kohen (1995) and Johnson & Wroe (2003) proposed an alternative hypothesis on the mainland extinction of the thylacine and devil: that the extinctions were due to an increase in human impact. Aboriginal people hunted thylacines. We know this partly from the diaries of George Augustus Robinson (Plomley 1966), who in the early years of British settlement in Tasmania travelled the whole island attempting to contact all Aborigines still leading a free existence, and made many observations on their way of life. He refers to thylacines only occasionally, but his observations and conversations with the people suggested that they sometimes killed and ate them; one group boasted that they had 'speared plenty'. (One incident also suggests that Tasmanian Aborigines held thylacines in awe. Robinson's party killed a female thylacine on 18 June 1834 and left the body lying in the open. Several days of poor weather followed, which the Aborigines in the group blamed on 'the circumstance of the carcase of the hyaena being left exposed on the ground and the natives wondered that I had not told the white men to have made a hut to cover the bones, which they do themselves, make a little house'. Plomley (1966) likened this 'little house' to the canopy that was traditionally erected over the ashes of a human cremation.

Hunting of thylacines is illustrated in northern Australian rock art. Two 'dynamic figure' paintings show thylacines being carried by people: in one a man holds a thylacine by the neck, and in the other it is slung over his shoulder (Chaloupka 1993). Two somewhat younger paintings are even more explicit. One shows a man lunging towards the tail of a thylacine, with a row of dashes coming out of his mouth possibly signifying a shouted exclamation (Chaloupka 1993). The other shows a thylacine with a three-pronged spear in its back (Plate 17).

The natural abundance of thylacines was probably even lower than that of dingoes. Hunting by people might therefore have had a large impact on their populations, and this impact would have increased as the human population grew through the Holocene. Furthermore, thylacine populations might have been stressed by reductions in populations of their prey caused by increases in the hunting efficiency of people and their numbers. The effects of people on the abundance of the thylacine's prey may well have been very large (see below), thus putting pressure on both dingoes and thylacines. Dingoes, being able to resort to other foods and to scavenging, and receiving the demographic subsidy of camp life, lived through this pressure. Thylacines, having neither of these advantages, did not.

Some particular innovations in hunting technology may have been critical in the tightening of ecological pressure on thylacines. The hard dates on disappearance of thylacines in southern Australia follow soon after the establishment of stone spear points and backed flakes throughout the mainland. Stone-tipped projectiles kill large animals more efficiently than do wholly organic spears, and they are associated with large-mammal hunting worldwide (Boeda *et al.* 1999; Ellis 1997). Their appearance may have marked a stepping up of the hunting pressure on large mammals, including thylacines. Backed flakes are thought to have been arranged in rows as barbs on hunting spears, to increase the size of a spear wound and the consequent blood loss. In cases where making a kill involved tracking a wounded animal, barbs would have increased success by making the wound more debilitating.

But there is one piece of technology that may ultimately explain more about the decline and fall of the thylacine on mainland Australia than any other single factor. The spear-thrower increases the range, accuracy and stopping power of a spear, and is almost essential for spearing animals at long range. Thylacines were naturally rare, and no doubt they were alert and wary of people, so hunters probably had very few opportunities to kill them with thrusting spears or standard hand spears. But with the appearance of the spear-thrower a man sighting a thylacine at some distance may have had a fair chance of bringing it down, and with this change the vulnerability of thylacines to human hunters suddenly increased.

Davidson (1936) argued from a comparative analysis of spear-thrower and spear types that the spear-thrower appeared first in northern Australia and spread south. Just when this happened is unclear because there is no archaeological record of spear-throwers, except for the negative evidence that a 10-kyr-old cache of wooden implements recovered from a waterlogged swamp in the far southeast of South Australia included no spear-throwers among a total of twenty-five items (Luebbers 1975). However, the development of the spear-thrower in Australia can be traced through rock art. In the Kimberley, spear-throwers first appear in the 'clothes peg figure' period which, as noted above, may be 20 kyr or more old. Subsequent art styles of the Kimberley record a continuous diversification of spear-thrower types and associated spear designs (Walsh & Morwood 1999). The 'clothes peg figure' period is also the last art style of the Kimberley to include depictions of thylacines, and it is possible that thylacines became increasingly rare in paintings of this period (Walsh personal communication). This is one piece of evidence for a correlation between the use of the spear-thrower and extinction of the thylacine. Another comes from the distribution of spear-thrower use throughout Australia. Spear-throwers were used throughout the mainland, but their abundance relative to hand spears was least in parts of southern, southwestern and coastal western Australia (Davidson 1936).

If the spearthrower remained rare in southern Australia, this might explain why thylacines persisted longest in the south. Most obviously, the spear-thrower never reached Tasmania, and this may be enough to explain why thylacines survived there until recently. More generally, many of the features that indicate late Holocene intensification on the mainland are absent in Tasmania, and human impact on the environment probably did not increase over the last few thousand years as it did on the mainland (see Chapter 8).

There is evidence from rock art that devils too were hunted by people on the mainland. A human burial from western New South Wales contains a necklace consisting of 178 devil incisors, from at least 47 individuals (Flood 1999). This is an indication that devils may sometimes have been killed in very large numbers. The pattern of decline of devils paralleled that of thylacines, although it might have occurred somewhat later. Devils would have been more abundant than thylacines, and they had higher reproductive rates. This would have made them more resilient to hunting by people, and could explain why they survived longer than thylacines (if that is what happened). As well as being hunted themselves, devils might have been affected by a reduction in carrion as a result of the declining abundance of large mammals.

It is possible that the extinction of the thylacine itself made life harder for devils. Thylacines in Tasmania reputedly ate only the soft tissue of their kills, and left large portions of a carcass behind. Devils, with their far greater ability to crush bone, were well equipped to finish the work of eating that the thylacines had begun. In this respect the two species were ecological partners, and a healthy population of thylacines would have been a boon for devils. By contrast, people carried the whole of their kills back to camp, where the remains would be scavenged by the resident dingoes. Australia's dominant predator of the last few thousand years was far less kind to the continent's most specialised scavenger than the thylacine had been.

To conclude, the evidence on what caused the extinction of the thylacine and devil is not yet strong enough to choose between the two main alternatives, dingoes and human impact. However, I do think it is possible to show that human impact provides a better fit to the facts than does the dingo hypothesis. A critical test will be to discover whether the thylacine really did go extinct in northern Australia before the dingo arrived. The evidence on this point might come from new fossil discoveries, but more probably from improvements in the dating of paintings of thylacines and dingoes respectively.

OTHER MAMMAL DECLINES

Lying behind the extinction of the thylacine and devil on mainland Australia was a general decline in the abundance of large mammals through the late Holocene as a result of heavy hunting by people. At least this is suggested by the fact that population densities of many species of mammals were low when the first Europeans arrived, but quickly increased following the effective cessation of Aboriginal hunting. Many early explorers and settlers commented on the scarcity of kangaroos through much of inland Australia (Barker & Caughley 1992; Flannery 1994; Rolls 1969). This was sometimes a matter of life and death. In 1873 the explorer Colonel Peter Warburton, on the point of starvation inland of Port Hedland in northwestern Australia, wrote that 'Trusting to game is a very precarious mode of getting food'. He suggested that efficient harvesting of wild food by the local Aborigines was at least partly to blame: 'The trees have all been carefully searched by the blacks for native honey, they have left none for us; could we find some it would be an indescribable treat' (quoted in Rolls 1969). Ernest Giles (1889) commented on a number of occasions that kangaroos and emus in central Australia were hard to shoot, being very wary as a result of heavy hunting by Aborigines.

Charles Sturt travelled through western New South Wales and southwestern Queensland in 1829 and 1845 and saw few kangaroos or emus. Only a few decades later pastoralists in these regions were alarmed by the rapid increase in kangaroos, which in some places were killed in their thousands. Rolls (1969) cites the astonishing figure of 80 000 kangaroos killed in one drive on Trinkey Station near Gunnedah. The New England region of New South Wales was first explored in 1812 and settled in the 1830s. One settler recalled that 'kangaroos were ... very scarce when white people first came to New England, and ... they increased very quickly between the thirties and the seventies'; in the 1880s, 10 000 were killed on one property in a year (Johnson & Jarman 1976). Western grey kangaroos increased at about the same time in southwest Western Australia (Abbott 2003) as did red kangaroos *Macropus rufus* and wallaroos *M. robustus* in central and western Australia following the establishment of European pastoralism (Newsome 1975).

An exception to this pattern was Tasmania, where the trend was reversed. Eastern grey kangaroos were initially plentiful around the first British settlements and they provided an important food source during the precarious early years of settlement, but they subsequently became rare as a result of destruction by Europeans and their dogs (Barker & Caughley 1990). The abundance of kangaroos in Tasmania in the early nineteenth century supports the case made earlier that, with their low populations and simple weapons, Tasmanian Aborigines had less impact on mammal populations during the late Holocene than did mainland people.

There are similar stories for koalas and tree kangaroos. Many early collectors had difficulty finding koalas in Victoria (Warneke 1978). A historical account from the lower Goulburn district (Parris 1948) states that koalas were initially unknown to European settlers, but began to appear in 1839 and were abundant by the 1860s. Parris believed that the increase was correlated with the decline of local Aborigines. Similar changes have been recorded for the Sydney region, where koalas expanded from dense forests and inaccessible sites into more open and flat country (Strahan &

Martin 1982). Koalas in less densely wooded habitats and easier terrain were, presumably, exposed to hunting, and there are many observations suggesting that they were a favoured prey (Martin & Handasyde 1999).

In north Queensland rainforests, tree kangaroos were initially very hard to find. Bennett's tree kangaroos *Dendrolagus bennettianus* were restricted to a few high-altitude sites not visited by hunters because of their spiritual significance (Martin 1996). However, with cessation of traditional hunting over the last fifty years they have extended their range into lowland forests and, very recently, into gallery forest along watercourses in savanna habitats (Martin personal communication). Tree kangaroos living in situations like these would have been very exposed to hunters in earlier times.

Smaller mammals also increased soon after European settlement. Bilbies *Macrotis lagotis* were not noticed by the first explorers to cross the Riverina district of New South Wales, but by 1877 a correspondent wrote in the *Riverina Grazier* that in places bilbies were so common that 'the greatest caution is necessary for horsemen lest one's horse's forelegs should go through the burrows' (Rolls 1969). In New England possums *Trichosurus vulpecula* and rat-kangaroos (probably the rufous bettong *Aepyprymnus rufescens*) were, like the large kangaroos, scarce in the 1830s but by the 1870s regarded as significant pests. Another account of a population explosion of possums at about the same time comes from the diaries of William Ross Munro (quoted by Kerle 2001), who bought a property in the St George district of southern Queensland in 1881, and watched brushtail possums increase from moderate numbers to 'millions': 'At night, on any night, they could be seen in clusters like fruit on a laden tree'. There were so many possums that they stripped the trees bare and denuded and befouled sheep pastures.

In southwest Western Australia possums, bilbies, brush-tailed bettongs *Bettongia penicillata* and burrowing bettongs *B. lesueur* all became abundant late in the nineteenth century and were pests of farm gardens, and there were also increases in abundance of brush wallabies *Macropus irma*, quokkas *Setonix brachyurus* and tammar wallabies *M. eugenii* (Abbott 2003). In the Bega district on the south coast of New South Wales, which was settled by Europeans in the 1830s, there were increases in populations of pademelons *Thylogale thetis*, rat-kangaroos, wallabies and bandicoots, as well as koalas and kangaroos, between about 1860 and 1900 (Lunney & Leary 1988).

The increases of kangaroos on mainland Australia are often attributed to habitat changes such as clearing of scrub to create pasture for sheep and cattle. This may have played a part, but Rolls (1969) argued that in western New South Wales kangaroos increased before extensive clearing of scrub. Mammal increases in southwestern Australia during the latter part of the nineteenth century cannot easily be explained by habitat modification, as land-clearing did not become extensive until after 1914 (Abbott 2003). Another possibility is that persecution of dingoes by European settlers allowed kangaroos to increase. We know too little about changes in the abundance of dingoes in the eighteenth and nineteenth centuries to judge this hypothesis, but again it seems that the kangaroo increase might have been too rapid to be explained as a secondary consequence of dingo control.

Bilbies, possums, koalas, bettongs and tree kangaroos are far less likely than large kangaroos to have been favoured by land-clearing and sheep-grazing, and the arboreal species were also relatively immune from dingoes. The fact that all increased in parallel implicates relief from Aboriginal predation as the main cause. The magnitude of the increases suggests that Aboriginal harvesting had been holding populations of these species well below their potential abundance. Probably this effect of people was relatively small about 10 kyr ago, but then increased as the human population grew, and especially from about 4 kyr ago as new hunting technologies were adopted. In some cases the reductions of prey populations may have set in quite recently. Stone hatchets were extremely useful for extracting animals from tree hollows or to fell small trees in which possums took refuge. In some parts of southern Australia possums were one of the major animal foods eaten by Aborigines, but the hatchets that were used to procure them spread into southern Australia only recently. In southwestern Australia people began using hatchets only within the last few hundred years before European arrival (Morwood & Trezise 1989).

THE TRANSITION TO *TERRA AUSTRALIS*

This history of Australian mammals has now reached the year 1788, when European settlement began. There had been huge changes to the continent's mammal fauna over the preceding 45 kyr, and I have argued that these are best accounted for by the effects of human hunting. The changes began with the extinction of the megafauna, which removed all very large vertebrates from the continent. The most significant aspect of this extinction for Australian ecosystems was probably that all large browsers suddenly vanished, causing ecosystem changes that we still do not understand. The size distribution of mammal communities was further squeezed by the dwarfing of the few large-bodied species that survived the extinction event. The contraction of Australia's mammal fauna continued in the Holocene with the mainland extinction of the remaining large mammal carnivores, the devil and thylacine. This loss of large carnivores was partly made good by the dingo, but there is a strong possibility that the numbers and ecological impact of dingoes were kept under increasingly tight control by people.

Within the last few thousand years, people emerged as the dominant predators on the continent. As resource use by people intensified on the mainland the surviving large and medium-sized mammals became steadily more scarce. The strange and wonderful mammal fauna that the first British settlers found in Australia was like a clockwork mechanism with many of its most important parts removed, and the others over-tightened. It was seriously out of equilibrium, and ready to fly apart.

Part III

EUROPEANS AND THEIR NEW MAMMALS

[The last 200 years]

By the Wonkonguroo the animal is called Yallara, and is evidently a widespread and well-known form … Its burrows are found only in the sandhills, never on the flats, and as the entrance is blocked with loose sand when the animal is within, they are never easy to locate, and in periods of wind are indicated only by a shallow dimple on the sloping surface of the dunes … From the entrance the burrow descends steeply for about two feet, then turns sharply, sometimes in a horizontal, sometimes in a vertical plane. On two occasions the distance from the entrance to the end was eight to ten feet in a straight line, and though there were several turns in both planes, the resultant course would not be a complete spiral …

The animals completely belied their delicate appearance by proving themselves fierce and intractable, and repulsed the most tactful attempts to handle them by repeated savage snapping bites and harsh hissing sounds, and one member of the party, who was persistent in his attentions, received a gash in the hand three quarters of an inch long from the canines of a male. On subsequently placing this male in a wire cage trap, already occupied by a large specimen of the pallyoora – the local pseudomys – the rat was immediately attacked and killed by the Yallara …

The stomachs of those which were dug out … contained large quantities of the skin and fur of rodents (but no bone fragments), seeds of a solanum(?), and some sand. No insect fragments could be made out.

Finlayson's (1935b) notes on the biology of the lesser bilby *Macrotis leucura*. These are the only records of the ecology and behaviour of this species, which went extinct in the central deserts *circa* 1960.

10

Mammal extinction in European Australia

THE FIRST TWO waves of extinction were similar in that they removed very discrete components of the fauna. In the first, this was large-bodied, slow-reproducing species from mainly open habitats; in the second, it was large carnivores. Whatever caused the third wave of recent extinctions also made a consistent distinction between survivors and casualties. To a very large extent, the species that vanished were medium-sized ground-dwelling species of dry habitats.

Some of the species that disappeared or declined drastically within the last two hundred years had distinct ecological roles, and with their passing Australian ecosystems lost some critical functions. This is an aspect of the recent extinctions that is rarely considered, so I spend some time exploring it at the end of this chapter. It is also one point where we can think optimistically about reversing the effects of past declines, because many of these ecologically important species still cling to existence as small remnant populations and could be restored to large areas of their original ranges.

THE EXTINCTION PATTERN

Mammal species that were living in Australia when Europeans arrived but are now extinct on the mainland are listed in Table 10.1. Eighteen species are completely extinct, and another nine have gone from the mainland but survive on islands (including Tasmania, which has three species recently extinct from the southeastern mainland).

BODY SIZE AND GEOGRAPHY

These recent extinctions were heavily concentrated on middle-sized species (Figure 10.1). Burbidge & Mckenzie (1989) identified a 'critical

Common name	Scientific name	Last record	Body mass (g)
Totally extinct			
Darling Downs hopping-mouse	*Notomys mordax*	1840s	?
Big-eared hopping-mouse	*Notomys macrotis*	1843	~ 60
White-footed tree-rat	*Conilurus albipes*	1845	200
Great hopping-mouse	*Notomys* undescribed sp.	1850 (?)	~50
Gould's mouse	*Pseudomys gouldii*	1857	~50
Broad-faced potoroo	*Potorous platyops*	1875	800
Eastern hare-wallaby	*Lagorchestes leporides*	1889	3 000
Short-tailed hopping-mouse	*Notomys amplus*	1896	~100
Long-tailed hopping-mouse	*Notomys longicaudatus*	1901	~100
Thylacine	*Thylacinus cynocephalus*	1930	30 000
Lesser stick-nest rat	*Leporilus apicalis*	1933	150
Toolache wallaby	*Macropus greyi*	1939	15 000
Desert rat-kangaroo	*Caloprymnus campestris*	1950s (?)	900
Pig-footed bandicoot	*Chaeropus ecaudatus*	1950s	200
Crescent nailtail wallaby	*Onychogalea lunata*	1956	3 500
Central hare-wallaby	*Lagorchestes asomatus*	1960	1 500
Desert bandicoot	*Perameles eremiana*	1960s	~200
Lesser bilby	*Macrotis leucura*	1960s	350
Extinct on the mainland, persists on islands			
Shark Bay mouse	*Pseudomys fieldi*	1895	45
Banded hare-wallaby	*Lagostrophus fasciatus*	1906	1 700
Tasmanian bettong	*Bettongia gaimardi*	1919	1 700
Western barred bandicoot	*Perameles bougainville*	1930s	220
Tasmanian pademelon	*Thylogale billardierii*	1930s	5 450
Greater stick-nest rat	*Leporillus conditor*	1938	350
Burrowing bettong	*Bettongia lesueur*	1960s	1 500
Eastern quoll	*Dasyurus viverrinus*	1966	1 100
Rufous hare-wallaby (mala)	*Lagorchestes hirsutus*	1991	1 265

Table 10.1 Mammals that have become extinct on mainland Australia since the arrival of Europeans, with date of last record in the wild and body mass.
Notes: From Abbott 2002; Burbidge *et al.* 1988; Lee 1995; Maxwell *et al.* 1996; Menkhorst 1995; Strahan 1995; Watts & Aslin 1981.

weight range', between 35 grams and 5.5 kilograms, within which species were highly vulnerable. The critical weight range takes in the larger rodents and the mid-sized marsupials. Cardillo & Bromham (2001) and Fisher *et al.* (2003) pointed out that while the statistical effect of body mass on extinction risk is strong at the lower end of the critical weight range, it is not quite so clear that the risk declined significantly for species above 5 kilograms (see also Johnson *et al.* 2002). This is because there are relatively few species of Australian mammals above 5 kilograms, and some of them (the thylacine and toolache wallaby) have gone extinct while

FIGURE 10.1 Extinction of mammals on mainland Australia since European settlement, showing the body-size distributions of living (black bars) and extinct (white bars) species for **(a)** marsupials and **(b)** rodents; and **(c)** the relationship between body mass and date of final extinction for all species.
Notes: The regression of mass on extinction date is significant, $F_{1,24} = 6.98$, $p < .05$. Species are included if extinct on mainland Australia (see Table 10.1).

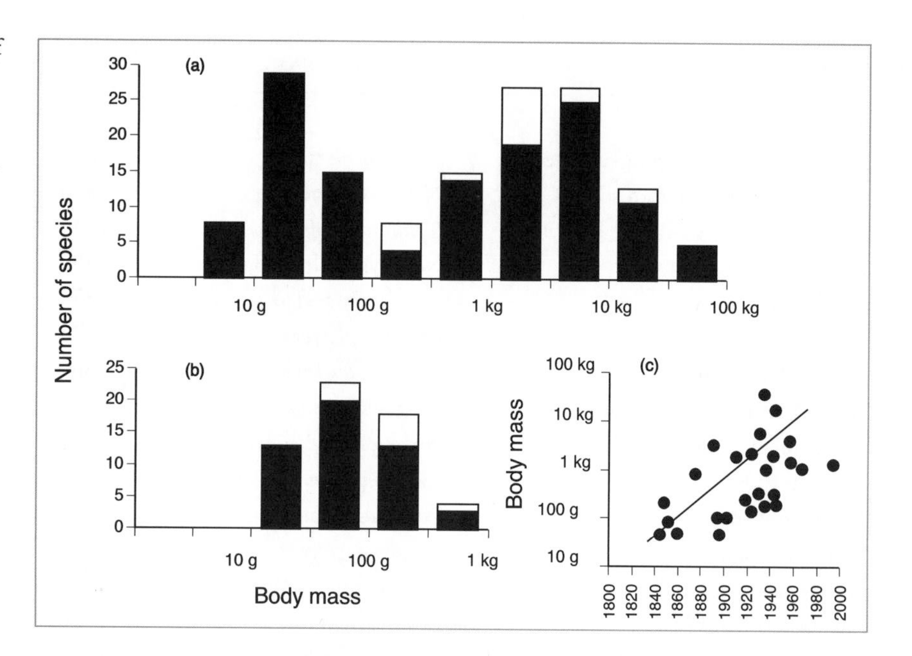

others, most notably the northern hairy-nosed wombat, are threatened with extinction. It is important to note, however, that these three large-bodied species all had very small populations when Europeans arrived. The phenomenon of widespread and abundant species collapsing to extinction within the last two hundred years is restricted to mid-sized mammals in Burbidge and Mckenzie's critical weight range.

The recent extinctions were concentrated in the drier inland regions of the southern half of Australia (Figure 10.2). Through most of semi-arid and arid Australia south of the tropics, *all* ground-dwelling mammals in the critical weight range have gone. The extinctions and declines did extend to higher rainfall areas close to the coasts in southern Australia, but in those regions they affected species that lived in the dryer and more open forest types. There have been no extinctions of species from tall closed forests, and none of arboreal species.

There is a third and less obvious feature of the selectivity of these recent extinctions: species that sheltered on the ground, rather than in trees (Plate 20) or in burrows or rock-piles, were most vulnerable (Smith & Quinn 1996; Burbidge & Manly 2002; McKenzie *et al.* 2005). For example, rock-rats and rock-wallabies (see Plate 19) occur through much of the territory that suffered severe mammal losses, and most species fall within the critical weight range but, although some species have declined to rarity, none has gone extinct. Among ground-dwelling rodents, burrowing species have fared better than species that did not burrow (Smith & Quinn 1996).

FOUR TOLLS OF THE BELL

The dates of the extinct species' last records in the wild on mainland Australia range from approximately 1840 to 1991 (Table 10.1). There was a trend for the smaller species to disappear first, with the very earliest casualties being rodents close to the lower boundary of the critical weight range (Figure 10.1c). Through the nineteenth and twentieth centuries the extinctions spread to larger species, such as the hare-wallabies, the toolache wallaby and the thylacine. There were several pulses of extinction, which affected very distinct components of the fauna. I think it might be possible to distinguish four of these, as follows (see Figure 10.2):

1840s and 1850s Five species of rodents from the southeast and southwest of Australia, including three species of hopping-mouse, were lost during this interval. All five originally occurred in dry open woodlands or heath, except for the great hopping-mouse, which evidently lived in grassland in the Flinders Ranges.

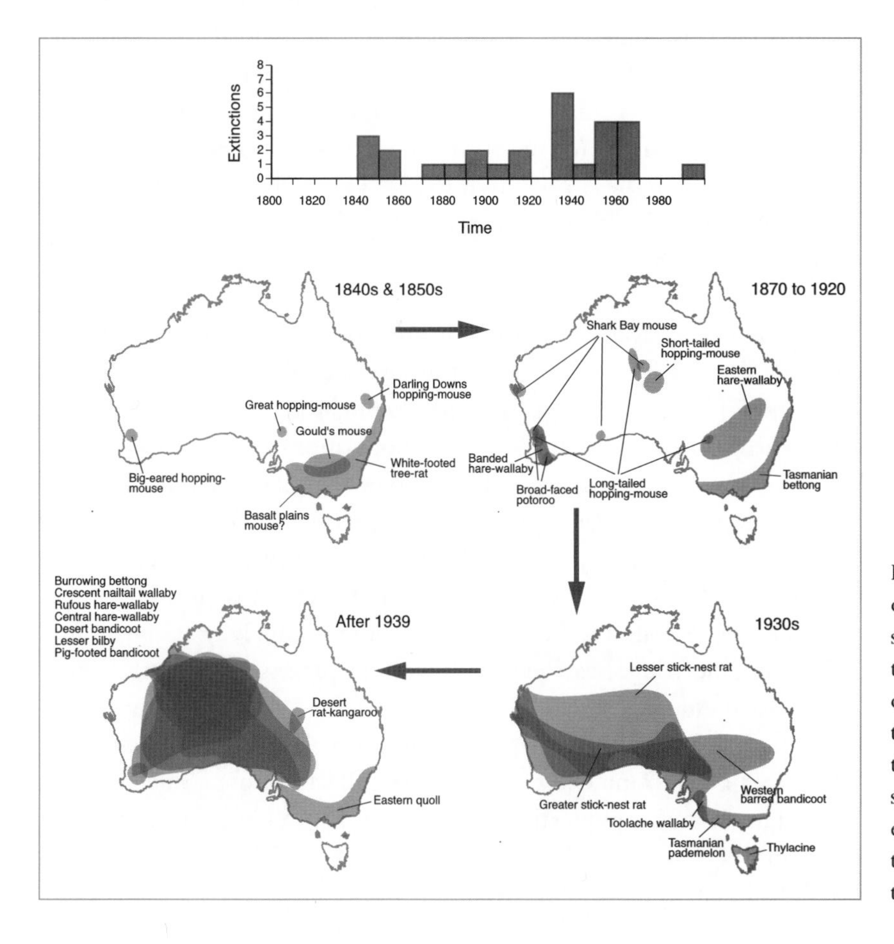

FIGURE 10.2 The pattern of extinctions since 1820, showing their number in ten-year intervals, and the original distributions on the mainland (and, for the thylacine, Tasmania) of species that finally went extinct in each of four time periods described in the text.

1880 to 1920 There were further rodent extinctions, but they affected species from more arid habitats across a much larger swathe of the inland. Two small kangaroos, the broad-faced potoroo and the banded hare-wallaby, disappeared from southwest Western Australia. In eastern Australia the declines of the Tasmanian bettong and especially of the eastern hare-wallaby were the first signals of what was to be the common pattern of extinction in the twentieth century – the utter collapse of widespread and abundant species from the middle and upper end of the critical weight range.

1930s A mixture of small wallabies, moderately large rodents and a small bandicoot disappeared across large areas of the southern inland. There were also two extinctions of large-bodied species from moderately high-rainfall habitats, the toolache wallaby and the thylacine.

After 1939 This late period saw the demise of a very distinct group: rat-kangaroos, bandicoots and small wallabies of the arid inland. The declines extended further to the north than they had earlier done. The species that went extinct disappeared first from the southern parts of their ranges and persisted longer in the north, in some cases until the 1960s. This wave of extinction was completed in 1991 with the loss of the last small wild population of the mainland form of the rufous hare-wallaby (or mala), but most of the decline of this animal took place in the 1940s and 1950s. The only exception to this pattern was the final disappearance of the eastern quoll from mainland Australia.

These four waves of extinction are not completely distinct, and defining them with reference to the last records of species has two problems. First, the last record of a species might be made long before its actual extinction, because small isolated populations can persist unknown for many years. This is dramatically (and blessedly) illustrated by the story of Gilbert's potoroo *Potorous gilbertii*. Discovered in 1840 by John Gilbert, John Gould's principal collector, Gilbert's potoroo was reported as very common in the Albany area of southwest Western Australia. Specimens continued to be collected from that area until the supply dried up around 1870. Guy Shortridge searched for it without success in the early 1900s, and it was presumed extinct. Then, in December 1994, Elizabeth Sinclair trapped a live animal in Two Peoples Bay Nature Reserve, only about 35 kilometres away from Gilbert's original collection site (Sinclair *et al.* 1996). The surviving population, of no more than 30 individuals living in less than 1000 hectares of dense *Melaleuca* thicket (Courtenay & Friend 2004), had gone unnoticed for over 160 years.

Second, the date of the final disappearance of a species does not necessarily represent the timing of its decline to rarity. Species that originally had small geographic ranges may have crashed to extinction quickly, but declines of some of the very widespread species were staged over long periods as animals disappeared successively from different portions of their original range, with most of the damage being done to their populations well before their final extinction. Figure 10.3 illustrates this for the chuditch (or western quoll) *Dasyurus geoffroii*, numbat *Myrmecobius fasciatus* and bilby *Macrotis lagotis*, all of which declined over many decades and now survive in small remnants of their originally wide distributions.

Despite these problems I believe this four-stage scheme does capture the real pattern of mammal decline in Australia. The reason it does so is that within any region the collapse of mammal populations was very rapid and complete, and happened at the same time for many different species. This means that the extinction of narrowly distributed species also indicates the regional declines of more widespread species. Many historical accounts testify to the suddenness and the synchrony of these regional collapses. A good example is the experience of Charles Hoy, who travelled around Australia collecting mammals for the American Museum of

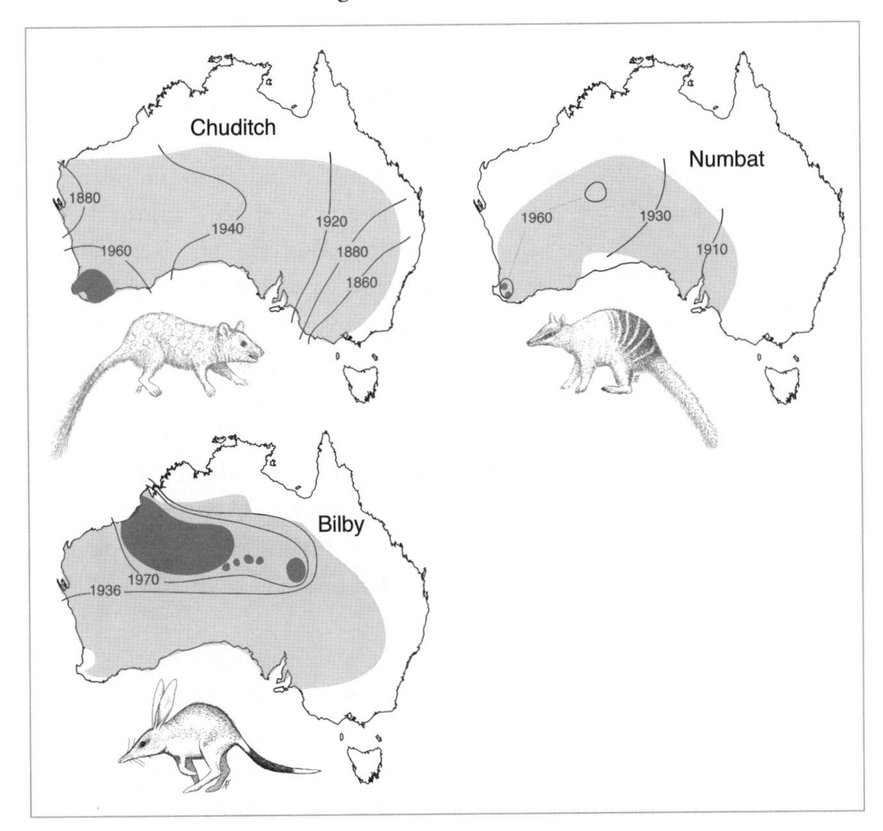

FIGURE 10.3 The decline in range over time of the chuditch *Dasyurus geoffroii* (Morris *et al.* 2003), numbat *Myrmecobius fasciatus* (Friend 1990) and bilby *Macrotis lagotis* (Southgate 1990). The contours represent approximate dates of disappearance from parts of the geographic range, where these are known. Dark shading represents distributions of surviving populations.

Natural History from 1919 to 1922 (Short & Calaby 2001). Hoy began with collecting trips in New South Wales and South Australia, where he was consistently frustrated by the rarity of native mammals and by the fact that wherever he went he seemed to be following the wake of recent catastrophic declines. He wrote:

> Australia certainly is a discouraging place for collecting. So many of the mammals are extinct and all the others nearly so. The sad fact of it is that, according to all reports, the mammals have only become scarce during the last few years. (Short & Calaby 2001)

At Port Lincoln on the Eyre Peninsula the magnitude and speed of the declines were illustrated by an old kangaroo hunter who told him that 'as late as two years ago he was sure of *at least* six or seven dozen wallaby skins a week while this year he hasn't even seen one!' Many other people provided similar accounts, and Hoy concluded with deep pessimism that practically all Australian mammals would be extinct within a few years. His collecting improved when he went to Western Australia, where at that time declines were less well advanced, and he was happy after his work on the intact mammal faunas of the tropical north.

The critical-weight-range mammals disappeared from the deserts of central Australia more than thirty years after Hoy's trip to Australia, but these losses may have been just as sudden. Finlayson (1961) travelled extensively through central Australia collecting information on mammals between 1931 and 1935, and again between 1950 and 1956. He recorded the mammal knowledge of Aboriginal people, and this approach was followed by later biologists, particularly Andrew Burbidge and Ken Johnson (Burbidge *et al.* 1988). These accounts from knowledgeable Aborigines collected over a period of about fifty years provide a large-scale picture of the declines, which shows that many species declined synchronously along an axis from south to north, culminating in final extinction around the mid-1950s and early 1960s. It was at about this time that most Aboriginal people moved off their traditional land and into permanent settlements. In 1979, when Ken Johnson began enlisting the help of the same Aboriginal men on mammal surveys, they were at first surprised and then saddened to discover that the animals that had been so familiar and important to them had almost completely gone (Johnson *et al.* 1996).

To summarise, the recent extinctions of mammals have the following characteristics: (1) they affected mainly medium-sized species from dry habitats in the southern parts of mainland Australia; (2) they were spread over a long interval of time, and they spread in waves from south to north and from east to west over the continent, affecting smaller species first and larger species later; and (3) within regions the process unfolded very quickly as populations of many species collapsed within short periods of time.

The declines affected not only the subset of species that crashed all the way to extinction, but also many others that disappeared from most of their original territory and now hang on in small remnant populations. About half of Australian marsupial species have had substantial declines in geographic range over the last two hundred years, and a third have disappeared from more than 50 per cent of their original geographic ranges (Maxwell *et al.* 1996). The majority of the species that have suffered major declines share the same characteristics of intermediate body size and occupation of dry and open habitats as the species that have gone extinct.

THE EXTINCT SPECIES

Five families of marsupials suffered extinctions – the Potoroidae (rat-kangaroos), Macropodidae (kangaroos and wallabies), Peramelidae (bandicoots and bilbies), Thylacinidae (thylacine) and Dasyuridae (carnivorous marsupials). The tally of extinctions accounts for just over 6 per cent of the marsupials that lived in Australia two hundred years ago, and 14 per cent of the rodents. No bats were lost from mainland Australia, and the two monotreme species were little affected.

Unfortunately we know little of the biology of most of the extinct species. Many disappeared before any European scientists made systematic studies of them. More disappointing still is that Aboriginal knowledge was ignored by most of the early European naturalists who did take an interest in mammals. This latter deficiency was partly corrected by some twentieth-century biologists, especially Hedley Finlayson who combined his own consummate field skills with a respect for Aboriginal knowledge and a systematic approach to its collection. He was also among the last European naturalists to see many of the species that are now extinct.

What we do know of the extinct species is summarised in Strahan (1995). The accounts below serve as a brief introduction in which I emphasise their most distinctive features, and their interactions with people.

RAT-KANGAROOS

Rat-kangaroos were once a familiar element of the mammal fauna in almost every part of Australia except the northern tropical savannas. They were, along with the hare-wallabies, a significant source of food and material for Aborigines and were hunted with almost unbelievable skill. They tend to be intensely curious animals and they became well known to European explorers and settlers as visitors to campfires and verandas. The three species that occurred in desert habitats – the burrowing bettong, brush-tailed bettong and desert rat-kangaroo (Plate 21) – all declined dramatically, as did the broad-faced potoroo (Plate 22), which lived in dry scrub fringing arid habitats in southwestern Australia.

The burrowing bettong (Plate 20) was one of Australia's most widespread species, being found over about two-thirds of the continent, but it now occurs only on three small islands off the Western Australian coast. It dug huge and complex burrow systems, comparable in scale to rabbit warrens. (Rabbits were in fact quick to move into them, and this may have been one factor that facilitated the rapid spread of rabbits through the southern arid zone.) The brush-tailed bettong may have been just as extensively distributed as the burrowing bettong, but it has contracted to a few small woodland patches in the far southwest of its range (where it is known as the 'woylie'). It was once common enough in South Australia that, according to Jones (1923–25), 'Twenty years ago the dealers in Adelaide did a great trade in selling them by the dozen at about ninepence a head for coursing on Sunday afternoons'. He also noted that though nimble they were not very fast and were easily caught by dogs, so presumably they did not provide very good sport.

The other bettongs and potoroos that live in more heavily wooded habitats in higher-rainfall regions near the coast have declined less severely, but all have suffered significant reductions in range (with the possible exception of the naturally rare long-footed potoroo *Potorous longipes*). The Tasmanian bettong was once widespread in dry open woodlands in southeastern Australia, but is extinct on the mainland. The most common rat-kangaroo now is the rufous bettong *Aepyprymnus rufescens*, but even this species has disappeared from much of the southern part of its range and it remains abundant only in the subtropics and tropics of eastern Australia. Rat-kangaroos eat fungi, fruit, roots, seeds and some invertebrates.

Aboriginal people in the desert country of the south of South Australia and adjacent areas of Western Australia have a word for a very small kangaroo ('wirlpa', 'weelba' and other variants) that does not match any species recorded alive within the last two hundred years (Tunbridge 1991). It is possible that this refers to the Nullarbor dwarf bettong *Bettongia pusilla*, which is known as a common animal from Holocene cave deposits in this region (McNamara 1997) and which on this evidence may have still been alive when Europeans arrived.

WIDE-AWAKE THE DANCING HARE, AND OTHER WALLABIES

The hare-wallabies are so called because they are about hare-sized, rest during the day in grass 'seats' (or, in the case of the rufous hare-wallaby, a short burrow) in open country, and start up at great speed if flushed. They were abundant throughout most of inland Australia in the nineteenth century. Only one species, the spectacled hare-wallaby *Lagorchestes conspicillatus* of northern Australia, remains widespread, but it has disappeared from the more southerly parts of its range on the mainland. The rufous hare-wallaby (or mala, Plate 34) and the banded hare-wallaby survive on Western Australian islands and in captive colonies, and they have been intensively studied.

We know little about the two species that are completely extinct. The eastern hare-wallaby (Plate 23) was reported as common, but it disappeared so quickly that little more is known of it than that it was an excellent jumper: John Gould had one bound clear over his head as it fled from his dogs. It was exceptionally fast, and in South Australia was known to settlers as the 'wide-awake' (Tunbridge 1991); the scientific name *Lagorchestes* means 'dancing hare'. The central hare-wallaby may never have been seen alive by more than one or two Europeans, and until Burbidge *et al.* (1988) began questioning Aboriginal people on it, the only recorded evidence of its existence was a single skull collected by the geological explorer Michael Terry in 1932. Aboriginal people recall that it lived on sand plains and sand dunes over a wide area of the western deserts of central Australia, until its extinction in about 1960. The crescent nailtail wallaby was also very widespread (Plate 24). It was first collected in Western Australia, but it also occurred over much of South Australia and the Northern Territory.

The toolache wallaby was the largest macropod to disappear over the last two hundred years and it was, by all accounts, one of the most beautiful of all wallabies (Plate 25). Finlayson (1927) glimpsed it in the wild and made close observations of the last surviving captive specimen. He compared it to the two most attractive living members of the genus *Macropus*, the whiptail *M. parryi* and western brush wallaby *M. irma*, commenting that it combined the most pleasing features of each, the 'elegance of form' of the whiptail blending with the 'striking yet harmonious character of ornamentation' of the brush wallaby. It was lightly built, with distinct black and white face markings, very slender forearms with white wrists and small black paws, black feet, and an otherwise fawn-grey body with distinct dark lateral bands over its rump. It lived in open grass and sedge country in the far southeast of South Australia, mainly in the vicinity of swamps, although it also penetrated semi-desert lands in the north of its distribution.

Because it was large, attractive and lived in open habitats it was very familiar to local people, who admired it not just for its beauty but also for its speed. In the early days of settlement, coursing toolaches with dogs was a popular sport. Finlayson was told that 'when flushed it almost at once attained a speed sufficiently great to extend the best of dogs, but this speed it continued to increase over a considerable distance, and even with greyhounds very long runs were the rule rather than the exception'. Its gait was distinctive – its body was held low and it made bounds of unequal length in a pattern often described as 'two short hops then a long one' – and it was capable of abrupt changes of direction at high speed.

The case of the toolache is one of the very few from the early twentieth century in which a concerted attempt was made to preserve a wild population of a species on the edge of extinction. By 1923 only about 14 toolaches remained in existence, at Konetta near the town of Robe. Professor Wood Jones of the University of Adelaide agitated for their protection, and the result was an attempt in 1923 and another in 1924 to capture animals and transfer them to a reserve on Kangaroo Island. Four animals were caught, but all died of stress and exhaustion. The publicity surrounding this ill-fated rescue mission sounded the doom of the remainder of the species:

> the realization of the great rarity of the wallaby roused the cupidity of an unscrupulous few, and that survivors of the 1924 attempt have been wantonly killed for the sake of the pelt as a trophy, is an assertion based on the admission of at least one of the slayers. (Finlayson 1927)

THE PASSING OF DESERT BANDICOOTS

The three extinct members of this group – the desert bandicoot, pig-footed bandicoot and lesser bilby – are all from the arid zone and were the smallest of the bandicoots and bilbies. Only the sparsest information is available on their natural history. The desert bandicoot (Plate 27) might have fed largely on ants and termites. The lesser bilby (Plate 28) may have been largely carnivorous (unlike other bandicoots, which are omnivores). The pig-footed bandicoot was the strangest of them all (Plate 26). Its limbs were long and slender, and its feet were very different from any other bandicoot's. The fourth toe of the hind foot was elongated and had a large terminal pad, while the other toes were reduced, so that the whole appendage resembled a tiny horse hoof. On the fore foot the second and third digits were equal in length and had large pads, while the other digits were reduced, and this foot looked like a pig's trotter. The animal stood on its toes and had an unusual erect, shuffling gait that reminded Krefft (1866) of 'a broken-down hack in a canter'. It lived in very open grasslands and sand plains, and it must have ambled about these habitats like a miniature horse. It appears to have been more herbivorous than the other bandicoots and bilbies. It is genetically very distinct from them and may belong in a family of its own (Westerman *et al.* 1999).

Two other bandicoots also occurred in arid and semi-arid Australia, the golden bandicoot *Isoodon auratus* in the north and the western barred bandicoot *Perameles bougainville* in the south. Both have retreated, the golden bandicoot to a reduced area of the tropical part of its range and the western barred bandicoot to Bernier and Dorre Islands in Shark Bay, Western Australia. The disappearance of the golden bandicoot in central Australia happened at about the same time that the pig-footed bandicoot and others went extinct; the western barred bandicoot declined from its more southerly distribution some decades earlier.

THE 'TASMANIAN TIGER'

The history and, as far as it is known, the ecology of the thylacine (Plate 29) have been reviewed by Guiler (1985) and Paddle (2000). As described in Chapter 9, the thylacine was a large specialist predator whose extinction in Tasmania was the culmination of a long decline that may have begun in northern Australia many thousands of years ago. It was evidently uncommon in the early days of European settlement, but held on for over one hundred years until its population collapsed about 1910. It had a reputation as a sheep killer, and no doubt it did kill and eat sheep from time to time, but the historical researches of both Guiler and Paddle make it clear that feral dogs were a much greater threat to Tasmanian sheep. Nonetheless, both the Tasmanian government and a large pastoral company, the Van Diemen's Land Company, introduced bounty schemes to encourage the destruction of thylacines. As I argue later, this incentive can explain why the thylacine population crashed when it did.

The thylacine is the only carnivorous marsupial to have gone extinct in the last two hundred years, but several other species, especially the chuditch and the eastern quoll, suffered major declines that mirrored those of the extinct species, and the eastern quoll is gone from the mainland.

RODENTS

Rodents suffered badly in the wake of European settlement, and all the species that went extinct belonged to the tribe Conilurini. The conilurines are the descendants of the first rodents to have entered Australia in the Pliocene. They are characteristic of semi-arid and arid habitats and they include the most distinctive and specialised of Australian rodents. Many species disappeared very quickly. Of the extinct species, only the white-footed tree-rat, Gould's mouse and the two stick-nest rats survived long enough to become familiar to European naturalists. The only evidence we have for the existence of most of the others is a few museum specimens (in the case of the Darling Downs hopping-mouse, a single skull). The 'great hopping-mouse' has still not been formally described.

The white-footed tree-rat (Plate 30) was a beautiful, large squirrel-like rodent that had a wide range in southeastern Australia. It made a mild nuisance of itself by raiding stores in the early settlement of Sydney, but apart from this it appears to have been uncommon and few Europeans recorded observations of it. Its habitat was probably dry open woodlands, and it extended into the semi-arid zone in western New South Wales (Lunney 2001). It denned in tree hollows, but we do not know how it foraged. The stick-nest rats (Plate 31) were similar in size to the rabbit-rat, but they lived in arid habitats and were named for the large piles of sticks they made. The explorer Ernest Giles described these structures:

> They form their nests with twigs and sticks to the height of four feet, the circumference being fifteen to twenty. The sticks are all lengths up to three feet, and up to an inch in diameter. Inside are chambers and galleries, while in the ground underneath are tunnels, which are carried to some distance from their citadel ... Their flesh is very good eating. (Giles 1889)

These large structures were occupied by (presumably) family groups. Stick-nest rats ate the leaves and fruits of herbs and small shrubs. Aborigines hunted them by setting fire to the nests and pulling the animals out ready-cooked. This seems too easy, but it appeared to have had little impact on stick-nest rat populations. Both species were widespread, and their nests can still be found in caves on the mainland even though there have been no rats to build them for more than sixty years. Some of these nests are thousands of years old (Pearson *et al.* 1999). The greater stick-nest rat survives on some South Australian islands where a translocation program is being implemented.

There were ten species of hopping-mice in Australia, five of which are now extinct. The extinct species were the largest members of the group, ranging from about 60 grams up to 100 grams or more, while the survivors weigh about 50 grams or less. The living species are very delicate and elegant animals; they eat seeds as well as insects and green plants and they construct extensive burrow systems that are occupied by groups of animals. Apart from their larger size, the other distinguishing characteristic of the extinct species is that their known geographic ranges were very small. This could be real, or it might be a function of inadequate collecting of species that crashed rapidly to extinction.

The long-tailed hopping-mouse (Plate 32) was collected from widely separated localities, and it is possible that when Europeans arrived it was continuously distributed through much of the intervening territory; evidence from subfossils supports this. John Gilbert commented that it burrowed in heavy clay soils, unlike most of the living species which occur mainly on sandy soils. The only other detail of its biology that he related was that it liked raisins. Gould's painting shows a plump animal that looked as much like a small rat-kangaroo as a large hopping-mouse.

Gould's mouse (Plate 33) and the Shark Bay mouse were relatively large members of the species-rich genus *Pseudomys*. Both were probably widespread through the inland but, like the extinct hopping-mice, their populations collapsed soon after European settlement. Another species, the undescribed 'basalt plains mouse' may have had a similar history, but it is known only from subfossils and could have been extinct before Europeans arrived.

MAMMAL DECLINE AND ECOSYSTEM DECAY

It is easy enough to imagine how the extinction of the marsupial megafauna might have disrupted Australian ecosystems. The repercussions of the decline and extinction of so many medium-sized mammals are bound to be more subtle, but they could be profound nonetheless. At the moment, it is possible to identify two major kinds of benefit that these mammals provided for ecosystems, and which have now been largely lost. First, many of them ate fruit and underground fungi, and in consequence they did valuable ecological service as dispersers of seeds and spores. Second, these species and others foraged by turning over the soil and some also dug burrows; these activities improved soil condition and helped plants to regenerate.

SPREADERS OF SPORES AND SEEDS

Fungi are a major part of the diet of many rat-kangaroos, bandicoots and rodents, especially species from forest and woodland habitats. These fungi are almost all subterranean (that is, their fruit-bodies form underground), and the mammals harvest them by digging. The largest fruit-bodies are about 5 centimetres across, and they resemble European truffles in that they are strongly aromatic, attractive to wild mammals, and provide a rich food reward for a mammal that follows the scent and unearths one. Among Australian mammals the bettongs *Bettongia* (with the exception of the burrowing bettong) and potoroos *Potorous* are truffle specialists (Plate 35). They eat truffles throughout the year and the contribution of fungi to their diet rises to 90 per cent or more when truffles are seasonally most abundant (Claridge & May 1994; Johnson 1994; McIlwee & Johnson 1998). These rat-kangaroos feed on a very large number of fungus species, up to fifty in some sites (Claridge & May 1994). They are expert at discovering truffles – professional mycologists can create lists of the subterranean fungi living in a stand of forest much more quickly and completely by identifying spores in the faeces of rat-kangaroos than by directly searching for fruit-bodies. Those bandicoots and rodents that feed on fungi typically eat fewer species than the rat-kangaroos, and do so only at certain times of the year.

The fungi that rat-kangaroos and other mammals eat are in almost all cases partners in a symbiotic association, known as a mycorrhiza, with plants. The nature of the association is that the fungus grows on the roots of the plant, explores the soil with its very fine hyphae, and extracts mineral nutrients and moisture which it passes back to the plant. In return, the fungus receives a supply of carbohydrate from its plant host. Mycorrhizal fungi are more efficient than plant roots at collecting nutrients and water from soil and, in the poor and dry soils typical of many Australian forest ecosystems, they are essential for plant growth. Plant groups like the eucalypts and acacias are heavily mycorrhizal. For a seedling of one of these plants to grow, it must be colonised by a mycorrhizal fungus. This can be achieved by the vegetative growth of fungal hyphae through the soil, but probably the more effective pathway is by dispersal of fungal spores.

This dispersal service is provided only by mammals that dig up and eat the spore-bearing fruit-bodies and then return the spores to the soil in faeces. The faeces of rat-kangaroos typically contain vast numbers of viable fungal spores of many different species (Plate 36). Glasshouse trials show that adding rat-kangaroo faeces to the soil in which a seedling has been planted is an excellent method of inoculating the plant with mycorrhizal fungi (Claridge *et al.* 1992). In the wild, rat-kangaroo faeces may be deposited many hundreds of metres from where the spores were ingested, and there is some evidence that passage through the digestive tract of a mammal stimulates spore germination (Johnson 1996). Simply by eating truffles, rat-kangaroos bathe forest ecosystems in a beneficial rain of viable and well-mixed fungal spores. The effectiveness of spore dispersal is enhanced if the faeces are buried by dung beetles, because this transports the spore load down into the root zone. Rat-kangaroos have commensal species of dung beetles that, clinging to the fur around the cloaca, ride under the mammal's tail until it defecates. The beetles then drop to the ground and immediately set to work burying the faeces (Wright 1997).

Rat-kangaroos complete a cycle that links fungi and plants in a mutually beneficial relationship and, with the help of specialised dung beetles, they ensure the establishment of mycorrhizal symbiosis on regenerating plants. The loss of rat-kangaroos could potentially have resulted in a decline in the abundance and diversity of fungi, a reduction in the formation of mycorrhizae, and ultimately a loss of productivity and diversity of plants (not to mention the disappearance of some very interesting dung beetles). One small-scale field experiment showed how strong these effects might be. Gehring *et al.* (2002) fenced terrestrial vertebrates (mainly large rodents) out of small plots on the rainforest floor in north Queensland. After three years they found a significant decline in the abundance and diversity of mycorrhizal fungi in the plots, and a reduction in the formation of mycorrhizal symbiosis on seedlings grown in soil taken from the plots. Whether these kinds of changes have happened at larger scales in forests and woodlands from which rat-kangaroos have disappeared is not known. To some extent the impact of the loss of rat-kangaroos would be tempered by the fact that bandicoots and rodents still occur in many of these areas, but these animals are probably less effective spore dispersers than the rat-kangaroos. It is quite possible that there have been subtle and unnoticed declines in the diversity and productivity of Australian forests as a result of the weakening of the ecological interactions between mammals, fungi and plants that followed declines of rat-kangaroos.

The rat-kangaroos and many other declined mammals also eat (or ate) fruit of small shrubs. Although it is not known for certain, there is a strong chance that fruit and seeds were a particularly important food for the larger conilurine rodents. Just as rat-kangaroos are important dispersal agents for fungal spores, they and rodents might have been important dispersers of the seeds of shrubs and small trees in the drier parts of Australia.

The storing of seeds in small caches is quite a common behaviour in rodents, and it sometimes involves burying seeds. This facilitates regeneration of the plants, because the mammal that buried the seeds does not always return to eat them before they germinate.

This behaviour has been recorded in recent times for the woylie in Western Australia. Christensen (1980) observed poison pea *Gastrolobium* regenerating on burnt ground very soon after fire, from seed transported and buried by woylies (and it is worth pointing out that these same woylies probably delivered a supply of fungal spores to those seedlings, by a different route). This observation suggests that seed-caching by woylies speeds up plant regeneration after fire. Woylies also cache seeds of quandong *Santalum acuminatum* and sandalwood *S. spicatum*, two species that produce large fruit clearly adapted for mammal dispersal. Murphy *et al.* (2006) showed that woylies 'scatter-hoard' seeds of these trees – that is, they carry them from where they find them and place them individually in small pits in an apparently random pattern, often returning to retrieve them but sometimes not.

In woodlands without woylies, sandalwood seeds accumulate on the soil surface at the base of the parent plant and have little hope of producing viable seedlings, but where woylies are present most seeds are moved away and buried. The difference is huge: Murphy *et al.* found an average of 60 seeds under the crown of each mature plant in a woodland that had lost its woylies, compared with only one under each plant in the presence of woylies. Sandalwood is failing to regenerate in much of Western Australia but, in the few woodland patches that have woylies, densities of seedlings and saplings are high, especially away from mature plants. The tree is commercially valuable because it produces aromatic oils used in making incense, and harvesting of sandalwood is an emerging industry in many parts of Australia and the Pacific. In Western Australia at least, this harvest will depend on the re-establishment of woylie populations.

ECOSYSTEM ENGINEERS

Rat-kangaroos collect truffles by digging to depths of as much as 20 centimetres. Good rat-kangaroo habitat is densely peppered with little pits and the associated spoil heaps created by the animals as they forage. The power of rat-kangaroo foraging as a mechanism of soil turnover was demonstrated by Garkaklis *et al.* (2004). They showed at a woodland site in Western Australia that individual woylies created diggings at a rate of between 38 and 114 per night, with each animal turning over an average of 4.8 tonnes of soil each year. The resident woylie population created between 5000 and 16 000 diggings per hectare per year, and turned over 3.2 tonnes of soil per hectare per year. Rat-kangaroo diggings alter soil characteristics in a number of ways. The soil is loosened; water infiltration is increased in and around the diggings; the diggings fill with organic matter; and dense mats of fungal hyphae grow through this matter as it decays (Garkaklis *et al.* 1998, 2000). The overall effect of the diggings is probably to increase the retention of moisture and organic matter, prevent soil compaction, and create fine-scale variation in topography and physical condition of soil.

Bandicoots and bilbies also forage by digging into the soil, and this behaviour may have been particularly important to the health of arid ecosystems. In the dunefields of the Arid Recovery reserve at Roxby Downs in South Australia, where bilbies and burrowing bettongs have recently been re-established, James (2004) measured a density of 6813 pits per hectare, about three times higher than in surrounding areas in which rabbits and goannas were the only vertebrates to feed by digging. Pits made by bilbies and bettongs had higher concentrations of moisture and nutrients than the surrounding soil, and the litter collected in the pits contained many seeds. James (2004) grew 1307 seedlings of at least 47 plant species out of the litter taken from 90 pits, compared with none from the same number of soil samples collected away from pits. In the absence of the marsupials, this organic matter with its nutrients and seed bank would mostly be swept away by sheet erosion.

Clearly, bilbies and bettongs have a very important role in providing the physical conditions needed for plant germination in such nutrient-depleted and dry environments; the effect of rabbit foraging pits is much smaller (James 2004). The subtle variations in conditions of moisture and nutrient availability that are created by disturbances like mammal diggings also provide a range of niches in which different species of plants regenerate, so they may play a part in maintaining plant species diversity. Some of the beneficial effects of foraging digs are also conferred by burrowing. For example, bilbies burrow in sandy soils and their burrows eventually slump and collapse, creating small depressions that accumulate water and organic matter. These become excellent sites for the establishment of deep-rooted grasses (Toby Piddocke, personal communication).

So far, research on the ecosystem effects of bandicoots, rat-kangaroos and rodents has just scratched the surface (so to speak) of this important topic. For many extinct species, we will never be able to do more than guess at how they interacted with other plants and animals, and how these interactions affected ecosystem functioning. But, we know enough to be sure that the survivors in these groups are important keystone species and ecological engineers in woodlands and arid habitats. They facilitate dispersal and establishment of a wide range of other species; they improve soil condition and retention of nutrients and moisture; and by doing this they doubtless help to maintain plant cover and prevent erosion. The restoration of species like the bilby to their original distributions would have the double benefit of increasing the population health of the species and of restoring their lost contribution to ecosystem health. But before this can be done, we need to know what caused those species to decline so catastrophically in the first place.

11

What caused the recent extinctions?

THE DECLINES AND extinctions described in the last chapter belong in a chaotic period of the recent environmental history of Australia, when Aboriginal land management was being replaced by European settlement and many things were changing. Because the declines took place within a short time after European settlement in any region, there were few or no naturalists on hand to observe them; and by the time governments and scientific institutions were in a state to properly organise research, many mammals had already disappeared. Hence the confusion of hypotheses on what caused the extinctions. They have variously been blamed on disease, on competition from introduced herbivores (especially sheep and rabbits), on loss of habitats that had been maintained by Aboriginal use of fire, on habitat destruction and direct persecution by Europeans, on the interaction of these factors with drought, and on introduced predators, especially the red fox.

This is a long list of calamities, and over much of the continent native mammals suffered many of them simultaneously, or in quick succession. A sensible interpretation, therefore, is that mammal populations were overwhelmed by the combined pressure of changes that resulted directly or indirectly from the arrival of Europeans. The alternative view is that just one of these factors was critical in causing the extinctions, and that the extinct species would have been able to absorb the impacts of the rest were it not for that one lethal ingredient in the mixture.

Over the last ten years the evidence for this latter view – that a single factor was responsible for the great majority of the extinctions – has been steadily growing. It is becoming increasingly clear that predation from red foxes and, importantly, feral cats, played a decisive role in the great majority of the extinctions. To the extent that other factors contributed, it was mainly by amplifying the effects of predation by foxes and cats (Smith & Quinn 1996; Dickman 1996a).

This chapter and the next present the evidence for that conclusion. I begin by arguing that some plausible factors – disease, fire, habitat destruction and persecution by Europeans – had no direct role, other than in a small number of exceptional cases. I then review the evidence that foxes and cats were directly involved in most of the extinctions. Chapter 12 then considers why it was that these exotic predators had such destructive effects in Australia, and reviews the roles of rabbits and other introduced herbivores in that context.

SOME GOOD IDEAS THAT DON'T WORK

PLAGUE AND PESTILENCE

Fleay (1932) referred to a 'mysterious disease which annihilated many marsupials ... in the first years of this century'. Other naturalists from the early twentieth century made similar comments. According to Le Souef and Burrell (1926) the koala, which was once 'extremely numerous', had become very scarce because of a disease 'which swept it off in millions in the years 1887–8–9, and from 1900 to 1903'. Jones (1923–25) suggested that an epidemic disease was responsible for a dramatic decline in the eastern quoll that began about 1900. Some contemporary accounts refer to symptoms of disease. Charles Hoy commented that the koala disease had caused 'a great bony growth on their heads' (Short & Calaby 2001). Lunney & Leary (1988) quote the recollection of an experienced mammal trapper from the Bega area, Mr Edgar Holzhauser, that he caught 'quite a lot of tiger cats [spotted-tail quolls *Dasyurus maculatus*] ... in those early days' but 'around 1928 they got a severe mange and died right out'. Another observer recalled that as the koala population declined between 1905 and 1909, koalas were seen looking 'very sick and dejected ... before they were found in hundreds dead at the foot of trees'. (Both koalas and spotted-tail quolls are still present in the area, but are rare.)

In other cases no sick animals were seen, but populations declined so drastically and with so little apparent cause that a savage disease epidemic seems a plausible explanation. For example, Mr William Munro described the sudden end of a population explosion of possums in the St George area in the early 1880s, when possums swarmed over sheep pastures like rabbits. After a winter of heavy rain, the possums disappeared:

> Those millions and millions of possums just were not there after the flood. No man saw them die. No one knows what afflicted them. Doubtless every hollow tree served as a vault in which the crumbling bones may be found to this day ... As an indication of how complete was

> the eradication, it can be mentioned that Roy, my eldest son, born in '86 asked me during '96 to explain what a possum was like. He had never seen one. The possums came back gradually. (quoted by Kerle 2001)

Europeans brought many mammal species to Australia and some of them became widespread very quickly, especially the feral cat (see below) and the house mouse. It is possible that these mammals introduced diseases to native species in time to explain the epidemics in the late nineteenth and early twentieth centuries. There are several candidate pathogens, but the one likely to have been most significant is the protozoan parasite *Toxoplasma gondii*, which causes toxoplasmosis (Dickman 1996b; Freeland 1994). This parasite lives in the small intestine of the cat and its infective stages are shed in cat faeces. They can survive for long periods outside a host and may be accidentally ingested by herbivorous mammals, or ingested by earthworms and insect larvae then picked up by invertebrate-feeding mammals. *T. gondii* damages the brains and nervous systems of these intermediate mammalian hosts, causing blindness, disorientation, lethargy and loss of coordination. Toxoplasmosis can be directly fatal, but it also increases its victims' vulnerability to predators – this is in the parasite's interest, because it completes its life cycle if the predator is a cat.

The disease has been found in a wide range of native mammals, but especially in small macropods, bandicoots and rodents. Freeland (1994) pointed out that these are the mammal groups that suffered most extinctions, and he argued that toxoplasmosis or other analogous diseases were largely to blame. Less well-known diseases may have been introduced by house mice to native rodents, by rabbits, and by foxes. These are all species that reached very high abundance soon after they became established in Australia and could therefore have spread parasites and disease on a huge scale.

There is no doubt that some diseases crossed from introduced to native mammal species in Australia, and the historical evidence of disease epidemics causing population declines is reasonably persuasive. However, the disease hypothesis does not fit the observed extinctions. There are four problems with it.

First, there is little evidence that disease affected those species that declined to extinction. The historical accounts of disease epidemics refer to species such as koalas, possums and spotted-tail quolls, which remain reasonably widespread. Many central Australian mammals declined between the 1930s and 1950s. Aboriginal people knew them intimately, but in none of their recollections of those species is there any reference to sick or dying animals. As far as we can tell, they remained healthy as their populations disappeared. Toxoplasmosis may have high prevalence in some mammal populations within the critical weight range – Freeland (1994) cites figures of 40 per cent for the southern brown bandicoot *Isoodon obesulus* and 41 per cent for the Tasmanian pademelon – but these populations remain abundant. Thus we have evidence of high disease burdens in populations that remain large, and little evidence for disease in populations that declined to extinction.

The one exception to this picture is the thylacine. Both Guiler (1985) and Paddle (2000) noted many observations from around 1900 of sick and disoriented thylacines, when reports of mange also became common. During this period thylacines were being captured alive in the wild and transported to zoos in quite large numbers. Paddle (personal communication) has examined zoo records for these animals and found that their longevity in captivity declined after 1897, suggesting that increasing numbers of animals brought in from the wild were seriously ill. This is the best evidence we have of a significant demographic effect of disease on a population that was soon to go extinct (but, as I argue below, thylacines would probably have gone extinct even if they had remained healthy). The eastern quoll and Tasmanian devil also declined to rarity in Tasmania in the first quarter of the twentieth century and this is usually blamed on disease, perhaps the same one that afflicted thylacines (Green 1973), but populations of these species have since recovered.

Second, it is not clear why the impacts of disease should have been concentrated in the drier inland habitats of southern Australia, where the major declines and extinctions took place. The introduced species most likely to have triggered the disease epidemics of the late nineteenth and early twentieth century are the house mouse and the feral cat. Feral cats occurred throughout Australia, including the tropical north where there were no population declines like those in the south. If feral cats transmitted a fatal disease to native mammals, why were the effects not as widespread as the cats themselves? The transmission of toxoplasmosis is best favoured in high-rainfall areas, but mammal declines were greatest in the dry inland (Dickman 1996b).

Third, if disease was a cause of decline of species that are now threatened with extinction, and if the same pathogens are still being circulated through the environment by introduced mammals, then disease should re-emerge as a problem for the reintroduction of declined mammals to their original habitat. Many such reintroductions have been attempted and most have failed, but never because the animals sickened and died. Their typical fate has been to thrive until eaten by introduced predators (Short 2004; Short *et al.* 1992).

Fourth, the fact that reported disease epidemics happened during the early 1900s is odd. This is too early to be a consequence of diseases introduced by rabbits and foxes, which had not at that time become widespread, and seems too late to be explained by cats and house mice, which had been present since the 1850s and earlier. If marsupials had been highly susceptible to diseases introduced by these species we should have seen a pattern of disease spread that followed the invasion front of those host species, but we do not. Perhaps the conditions that allowed these disease epidemics to take hold were created by the massive increases in population density in species like possums and koalas that were, in turn, caused by release from Aboriginal predation. There could be a very interesting story here, but it is separate from the problem of explaining extinction, in which the weight of evidence is very much against disease as a significant contributing factor.

THE CHOICE OF FIRE OR FIRE

Aboriginal people used fire in a great variety of ways and for many purposes, but a general feature of their fire management is that it involved frequent low-intensity burns. This created a mosaic of habitat patches in different stages of recovery from fire: mature patches of spinifex, for example, would be found alongside recent burns in which a variety of other small plants could regenerate. Bolton & Latz (1978), Johnson *et al.* (1989) and Johnson & Roff (1982) argued that this fine-scale patterning of habitat was very favourable to mammals in the critical weight range, because it provided a wide range of foods and sheltering sites in close proximity to one another. The arrangement also worked as an effective system of fire breaks that impeded destructive wildfires.

When Aborigines moved off their land and into permanent settlements this fine habitat mosaic disappeared as vegetation was allowed to mature over large areas, which in turn meant that wildfires could sweep through vast tracts of country. In the immediate aftermath of a severe fire, mammals would lose both their food and shelter and would die (if they survived the fire itself). Therefore changed fire regimes caused population extinctions. How had those mammals survived before people started burning the country? Flannery's (1990, 1994) answer to this question is that in pre-human times browsing by giant marsupials had maintained something like the same vegetation pattern, and kept fuel loads generally low. After the megafauna went extinct, Aboriginal burning took over as the main factor regulating vegetation and preventing wildfires.

This hypothesis could be very useful in explaining extinctions in places like the northern parts of the central deserts, which were not occupied by Europeans or rabbits, had few foxes, and were largely untouched by introduced livestock. On the other hand, the pattern of fire and vegetation in those areas had been strongly influenced by Aboriginal burning and must have changed when that burning ceased. Mammal declines in the central deserts followed Aboriginal departure in a consistent pattern. For example, the disappearance of the chuditch in central Australia began in the Hermannsburg area, coinciding with the relatively early settlement of the Aranda people around the Lutheran mission there (Johnson & Roff 1982). In other deserts areas where Aboriginal occupation persisted longer, so did the chuditch.

The complete version of the fire hypothesis developed by Flannery proposes that Australian vegetation has been regulated by Aboriginal burning ever since human arrival (or, more precisely, since the megafauna extinction). This is its main flaw. As I argued in Chapter 8, Aboriginal management of vegetation by fire may have a much shorter history than that, and it probably reached its present form only within the last few thousand years. This is certainly true of the central deserts, where the fire hypothesis would seem to be most useful. Although the early dates on sites like Puritjarra and Kulpi Mara (see Chapter 4) show that people were in central Australia by close to 40 kyr ago, this was before the onset of fully arid conditions. Much of the arid centre was abandoned in the approach to the glacial maximum, and was extensively reoccupied only from about 4000 years ago. Occupation intensity has increased dramatically since, especially within the last 2000 years.

This means that for at least 20 kyr the vegetation of vast tracts of the centre was affected neither by mega-herbivores nor by Aboriginal fire management. If one or the other of these factors was essential to maintain habitat for medium-sized mammals, how did they survive in the absence of both? The most sensible interpretation is that these mammals were actually very resilient to changes in vegetation and fire regimes, and it was something else that made them go extinct in the recent past.

There are other problems with the fire hypothesis. Mammal decline might have followed Aboriginal departure in some places, but elsewhere there is evidence that many mammals increased soon after Aborigines left their land (see pages 162–5), and did not go into decline until much later. This implies either that the impact of changed fire regimes on the ecology of those species was not significant, or that it was overridden by the effect of reduced hunting. The bilby remained widespread in southwestern Western Australia even as fire was being used on a large scale to clear vegetation for farming (Abbott 2001). In central Australia the distribution of surviving bilby populations is apparently unrelated to fire history. Rather, substrate type determines where bilbies have been able to persist – they are found along creek lines and in rocky habitat, but not in open sandplain and dunefield areas (Allan & Southgate 2002). Likewise, the quokka *Setonix brachyurus* remains abundant on Rottnest Island off the southwestern Australian coast even though the vegetation of the island has changed from woodlands and scrub to low heath as a result of changed fire regimes (Johnson *et al.* 1989).

At the heart of the fire hypothesis is the idea that the mammals that declined depended on a fine-grained mosaic of vegetation, as created by recurring disturbances like fire. Short & Turner (1994) tested this by comparing populations of three mammal species – the brushtail possum, golden bandicoot and burrowing bettong – in areas of uniform and complex spinifex vegetation on Barrow Island, and found no effect of vegetation patterning on their distribution and abundance. Smith & Quin (1996) point out that the mammal species most sensitive to fire regimes are some small *Pseudomys*, which remain reasonably widespread.

To summarise, changed fire regimes seem to have made little if any direct contribution to recent mammal extinctions. The idea does remain attractive for the deserts because it explains why some mammals declined there as Aborigines left but, as I will show later, there is another explanation for this with stronger support. This is not to say that fire is not an important tool in the management of habitat for mammals and many other species; it certainly is, but we must look elsewhere to explain the disappearance of critical-weight-range mammals.

KILLING AND CLEARING

Neither direct persecution nor habitat destruction can explain much about the pattern of extinction, except in isolated cases. The most remarkable feature of Australia's mammal extinctions is that they were at their worst in the parts of the country where human population was least, and in some cases where it was negligible. Mammals disappeared from most of the Tanami Desert at a time when few Europeans had even visited it. A large proportion of the country that lost its mammals has never been cleared, and in other areas mammals disappeared before clearing was significant. The one case where habitat clearing might have been critical was the toolache wallaby, which lived on what was to become agricultural land in a small part of southeastern South Australia. By 1923, when it was on the edge of extinction, most of the toolache's habitat had been

FIGURE 11.1 The thylacine or Tasmanian tiger *Thylacinus cynocephalus.* (photograph by David Fleay, reproduced with permission of the Archives Office of Tasmania)

destroyed by draining of swamps and clearing of vegetation for crops (Robinson & Young 1983).

The other species that suffered heavy persecution at the hands of Europeans was the thylacine (Figure 11.1). From the time the British arrived in Tasmania they regarded the thylacine with a mixture of fear, resentment and blame, although there was little reasonable basis for any of these feelings. As noted earlier, thylacines were powerful predators and they occasionally killed sheep, but compared with feral dogs they were a minor threat to Tasmania's sheep and they very rarely, if ever, attacked people. Perhaps the rather dark and grim thylacine embodied the settlers' anxieties about their new home, and they felt a deep need to be rid of it. Or maybe, as Paddle (2000) suggests, the thylacine was made a scapegoat for the hardships that the wool growers experienced (which were mostly due their own mismanagement).

Whatever the motive, the common response of the British to thylacines was to kill them. The Van Diemen's Land Company, which had very large pastoral holdings in central and northwestern Tasmania, encouraged this attitude with a bounty on thylacines from 1830, and in 1887 the colonial government was persuaded by a lobby of sheep graziers to introduce a bounty scheme for the whole island (the bounty was set at one pound for an adult thylacine, ten shillings for a juvenile). The number of bounty payments made each year provide us with a record of the history of persecution of the thylacine and, indirectly, of trends in the thylacine population.

Figure 11.2 shows bounty payments on thylacines killed on the Van Diemen's Land Company's Woolnorth Station in the far northwest of Tasmania from 1874 to 1906. These figures come from detailed diaries maintained by the Woolnorth managers, and are probably accurate (Guiler 1985). From 1874 to 1884 up to 10 thylacines were killed each year. The property at that time covered 40 500 hectares. We do not know how abundant thylacines were, but historical accounts suggest that adult pairs moved through exclusive territories of at least 5500 hectares (Guiler 1985, Paddle 2000). This translates to a population of no more than 14 adult thylacines in an area the size of Woolnorth. Females probably reared two young per year, not all of which would have survived to independence, so the Woolnorth population might perhaps have produced ten juveniles each year in good conditions.

The harvest levels sustained by the Woolnorth thylacines therefore seem high enough either to have removed all the population's recruits, or to have killed all adults within a year or two. This is far above a sustainable harvest rate, and yet thylacines were killed at Woolnorth over a period of more than seventy years, and the highest annual kill was 19 in the year

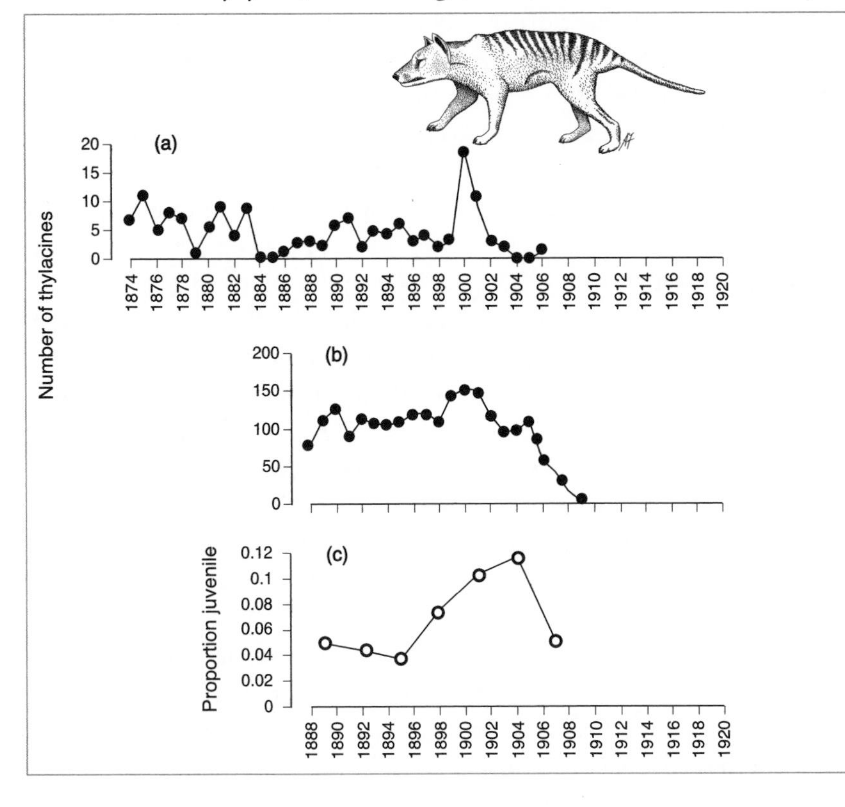

FIGURE 11.2 Number of thylacines **(a)** killed each year at Woolnorth Station in northwestern Tasmania and **(b)** presented for bounty payments under the government bounty scheme throughout Tasmania; **(c)** the proportion of juvenile thylacines among animals presented for bounty throughout Tasmania.

1900. This can only be explained by immigration. As thylacines were killed on Woolnorth others must have moved in from elsewhere, to be killed in turn. There was an underlying decline in the annual kill on Woolnorth, which is clear if the unusually high number in 1900 is excepted (when the regression of number killed on year is significant: $F_{1,30} = 7.72$, $p < .01$). This suggests that the population in the whole region around Woolnorth was gradually being depleted.

The trend in payments made under the government scheme for the whole of Tasmania shows some similarities to Woolnorth, particularly in their steep decline after 1905. There was also a high number of kills around 1900, but the difference is not so great as at Woolnorth. The government bounty totals are a minimum estimate of the numbers killed each year. Probably the true total was at least twice as high (Guiler 1985). Thylacines were also hunted for their skins and one report was that 'no fewer than 3482 skins were exported between 1878 and 1896' (Guiler 1985). If the overall harvest rate was anything like that at Woolnorth, thylacines should have declined rapidly to extinction. It is therefore interesting that the harvest remained approximately constant through the sixteen years from 1888 to 1904, when it suddenly crashed.

One interpretation of this crash is that the thylacine population was able to absorb the harvest that was being exacted, but that something else drove numbers down around 1905. Both Guiler and Paddle suggest that disease was this added critical factor and, as noted above, there is evidence of disease in the population at the time. Paddle believes that one effect of disease was to make thylacines easier to catch, and that this explains the increase in numbers killed around 1900 which, in turn, caused a quick decline to very low numbers. Maybe so, but there is a counter-argument, at least for Woolnorth Station where the spike in the number killed at 1900 is most marked. As I argued above, the animals killed at Woolnorth must have been mostly immigrants to the area. A sharp increase in the number killed must therefore reflect a large influx of immigrants, and it would be surprising to find animals moving long distances if they were seriously ill.

An alternative view is that with the institution of the government bounty, the effort devoted to thylacine hunting increased steadily and became more widespread. This caused a general population decline, but for a time the imposition of an increasing hunting effort on a declining population resulted in a roughly constant annual kill. As the population became smaller, a constant annual kill represented an increasing proportion of that population. This would have caused an accelerating decline, culminating in a final rush to disaster. Such an effect was produced by a model of thylacine harvest (Bulte *et al.* 2003).

There is another interesting trend in the bounty data, which provides some support for this interpretation: the proportion of juveniles in the kill increased through the period leading up to the crash. There could be two possible mechanisms behind this. First, the harvest might initially have selected for adults, either intentionally (because they were worth more than juveniles) or inadvertently (because, for example, adults were more likely to be caught in snares). As the population was depleted of mature adults and hunting effort increased, a rising proportion of the kill was juvenile. In this case, the shift in age-composition of the animals killed was a signal of a seriously overhunted population that was rapidly being made very small. Alternatively, a reduction in population size as a result of the kill might have provoked a density-dependent increase in breeding and juvenile survival. This would not necessarily have been a danger sign in itself, but the numbers suggest that the population had not come into balance with the harvest imposed on it between 1894 and 1904.

On either interpretation, the killing of thylacines was having a large impact on the demography of the population and was probably unsustainable. It is not surprising that the result was disastrous (or, according to the thinking of the time, satisfactory).

None of this evidence is decisive, but my feeling is that the thylacine crash of 1904–1910 can be explained by hunting alone, and that disease played only a marginal role. Thylacines persisted in very low numbers for at least another thirty years and were hunted in an opportunistic fashion until the last kill was made in 1930 (famously, six years before the species was given legal protection).

THE POWER OF INTRODUCED PREDATORS

THE RED FOX

Having discarded fire, disease, land clearing and human hunting as unimportant, at least in the majority of extinctions, we can turn to a suspect against whom the evidence is quite damning. There is no doubt that foxes played a major role in the declines and extinction of many Australian mammals.

The red fox *Vulpes vulpes* was introduced to Australia to provide sport for hunters. Hunting on the English model – men on horseback, in costume, following packs of hounds – was established in Australia as early as 1820 (Rolls 1969). To begin with, the victims were kangaroos and dingoes. But the hunt clubs yearned for more orthodox quarry, and

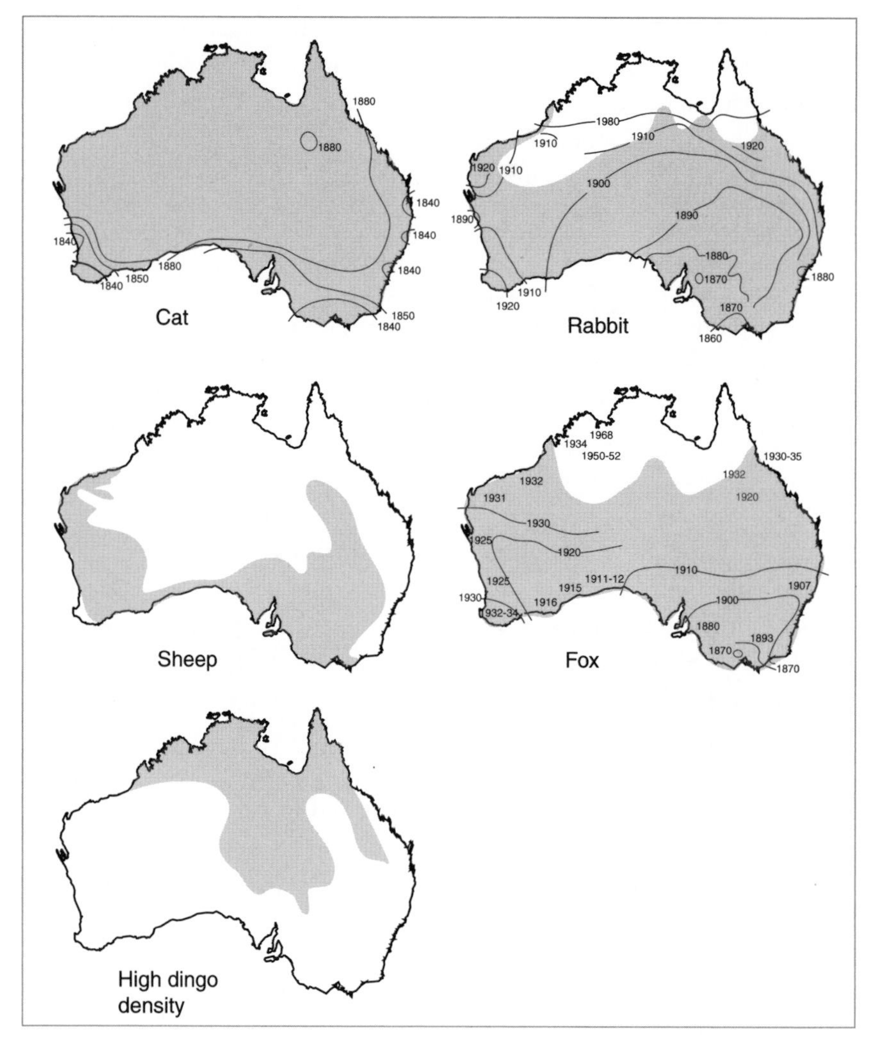

FIGURE 11.3 Historical spread of feral cats (Abbott 2002), rabbits (Williams *et al.* 1995) and foxes (Jarman 1986) over mainland Australia (the current distributions of these species are shaded), and the current distribution of sheep and high-density dingo populations (Fleming *et al.* 2001).
Notes: Fox and rabbit distributions from Saunders *et al.* (1995).

sometime around 1860 the first foxes were imported. Initially these were held captive, being released only to provide runs for the hunt clubs and then recaptured if they could be saved from the dogs at the last moment (deer were sometimes also used in this way).

Foxes were released in the wild near Ballarat and Geelong in southwestern Victoria in about 1870, and there was probably another release in South Australia soon after. The open plains of these regions, lightly wooded but with patches of cover, were ideal for foxes; rabbits and hares were already well established, and the foxes thrived. They spread rather slowly at first but the expansion gathered pace after 1900 (Figure 11.3). The population front crossed the Victorian border, moved through New South Wales and got to Queensland within the ten years from 1902 to 1912, travelling at up to 140 kilometres a year (Jarman 1986). Once foxes arrived in a location the

PLATE 31 Lesser stick-nest rat *Leporilus apicalis*, from John Gould's *Mammals of Australia* (1863).

PLATE 32 Long-tailed hopping-mouse *Notomys longicaudatus*, from John Gould's *Mammals of Australia* (1863).

PLATE 33 Gould's mouse *Pseudomys gouldii*, from John Gould's *Mammals of Australia* (1863).

PLATE 34 A captive mala (rufous hare-wallaby *Lagorchestes hirsutus*). The last wild population on mainland Australia disappeared in 1991, but the species survives on several islands off Western Australia and in captive colonies, where it breeds well. Attempts to reintroduce it to its mainland habitats during the 1990s were defeated by feral cats. (photograph by Ken Johnson)

population grew astonishingly fast, as indicated by bounty payments on fox scalps (foxes were quickly acknowledged as a major pest, and their destruction was encouraged). Numbers typically increased steeply to a peak within five to fifteen years, then dropped back to intermediate levels.

Foxes entered Western Australia by way of the southern Nullarbor Plain in about 1915, and within a few years they took over most of the southern semi-arid zone of the west. They occupied the forested environments of the southwest over a longer period, and their spread into the arid zone was also somewhat gradual. They reached their northern limit around 1965, but their distribution in the far north is sparse and patchy and they remain absent from large areas. Foxes did not establish in Tasmania, but there now appears to be a small population there, courtesy of a malicious introduction in about 1998.

The appearance of large numbers of foxes in a region was invariably followed by mammal declines. In New South Wales bounties were paid on a variety of marsupial species as well as on introduced pests like the fox; Short (1998) analysed bounty records for rat-kangaroos (including burrowing bettongs, rufous bettongs, brush-tailed bettongs, and possibly also Tasmanian bettongs and long-nosed potoroos). Rat-kangaroos were abundant in the years before foxes arrived, but as soon as foxes invaded they went into decline and were gone within ten years (Figure 11.4).

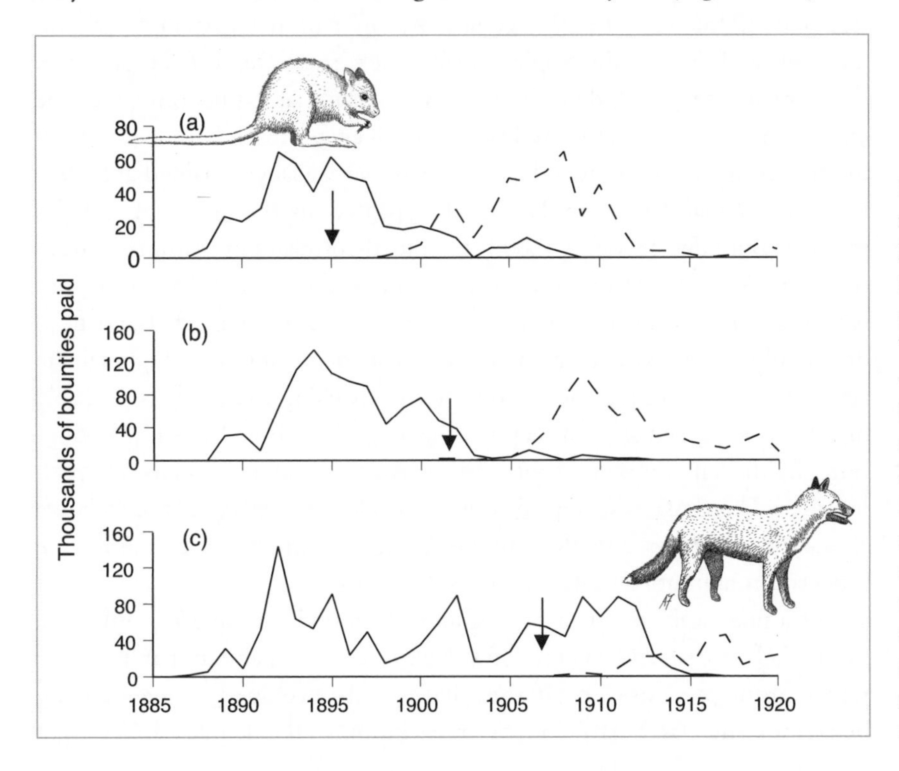

FIGURE 11.4 Increases in foxes and declines of rat-kangaroos in **(a)** southern, **(b)** central and **(c)** northern New South Wales. Foxes indicated by dashed lines, rat-kangaroos by solid lines. Vertical arrows indicate first arrival of foxes in each region. *Notes:* Values are number of bounty payments in each year, combining totals from the slopes and tablelands regions (from Short 1998), with numbers for foxes multiplied by five.

This sequence of events was very consistent, even though the arrivals of foxes in the regions analysed by Short were staged over fifteen years. It was certainly noticed at the time. Charles Hoy wrote that in the area of Tamworth wallabies and rat-kangaroos had been common, 'but soon after the advent of the foxes they disappeared so that now *Onychogalea fraenata* [the bridled nailtail wallaby] and *Aepyprymnus rufescens* [rufous bettong] are extinct and the others rapidly approaching extinction'. Rat-kangaroos in New South Wales survive only where foxes are scarce, and the burrowing bettong and Tasmanian bettong are completely extinct in that state.

The story was repeated elsewhere. Hoy dated the extinction of the tammar wallaby on the Eyre Peninsula to between 1915 and 1918, a little under ten years after the arrival of the fox (Short & Calaby 2001). The decline of the bilby in southern Australia followed the spread of the fox from east to west; it persisted in the far southwest until the 1930s, but vanished soon after the fox finally became established there around 1930 (Abbott 2001). Declines of other mammals, such as the quokka, also coincided with the relatively late arrival of foxes in the far southwest of Australia (Hayward 2002). Declines of marsupial carnivores, quolls (*Dasyurus* species) in particular, followed soon after the arrival of foxes throughout Australia (Jones *et al.* 2003).

The current distribution of the fox in Australia provides a close (though not perfect) match with the geography of mammal extinction. So, in Tasmania and most of the tropical north where there are no foxes or where foxes are still very rare, there have been no mammal extinctions (with the exception of the thylacine in Tasmania). Likewise, foxes have never been introduced to Kangaroo Island, and the tammar wallaby remained abundant there as its mainland populations disappeared in the wake of the fox invasion. Burbidge & Manly (2002) examined correlates of mammal extinctions on 176 islands around the coast of Australia, and found that extinctions were more likely on islands with foxes. Foxes are generally most abundant in dry, lightly vegetated habitats and least so in dense forest in high rainfall areas; also, their hunting success on ground-dwelling mammals is probably higher in open habitats. This can explain why there have been few extinctions in the tall forest of southeastern Australia, and so many in open habitats. There is a strong positive correlation between the relative abundance of foxes and the number of local extinctions of conilurine rodents throughout Australia (Smith & Quin 1996).

At a finer scale, remnant populations of ground-dwelling mammals are most likely to survive where they have access to habitat that provides refuge from predation, or if they have well-developed refuge-seeking behaviour such as burrowing or tree-climbing. Rock-piles and gorges provide mammals with shelter sites that limit access by foxes. Many Australian mammals, the rock-wallabies and rock-rats in particular, are specialised for living in rocky habitats. Foxes eat rock-wallaby when they can get it, and most rock-wallabies of southern Australia have declined since the fox invaded, but so far none has gone extinct. The risk of extinction of mammal populations on islands with introduced predators was reduced if the animals had access to rock-pile habitat; similarly, mammal species that either climbed or burrowed were more likely to survive with foxes (Burbidge & Manly 2002). Where foxes are abundant on the mainland, conilurine rodents that either burrowed or sheltered in rock-piles have suffered less decline than other species (Smith & Quin 1996).

Some woodlands in the southwest of South Australia have been partially protected from the impact of the fox by the presence of plants in the genus *Gastrolobium* ('poison pea'), the leaves of which contain the poison sodium monofluoroacetate (commonly known as 1080). Native mammals from the southwest have an evolved tolerance to 1080, while foxes are very sensitive to it. Where *Gastrolobium* is abundant foxes that eat native mammals are at risk of secondary poisoning, and this suppresses fox numbers. Woodlands with *Gastrolobium* therefore provided a refuge in which woylies, tammar wallabies, numbats and chuditches persisted as their populations disappeared elsewhere on the mainland (see Figure 10.3).

There is thus a wealth of comparative and historical evidence implicating foxes in the decline of many mammal species, especially rat-kangaroos and small wallabies. This is confirmed by recent evidence of a more experimental nature. The success of attempts to reintroduce threatened macropods to parts of their original range depends strongly on whether or not foxes are present. Short *et al.* (1992) reviewed twenty-five such attempts, and found that introductions to islands free of introduced predators were mostly (82 per cent) successful, while only 8 per cent of introductions succeeded if they were to mainland or island sites with predators. In some of these cases, released animals were followed by radio-tracking, causes of death were determined, and a high rate of predator kills was demonstrated.

Kinnear *et al.* (1998) reduced fox numbers around two isolated colonies of the black-footed rock-wallaby *Petrogale lateralis* in Western Australia, and compared them with another three at which foxes were not controlled. Black-footed rock-wallabies have disappeared from most of the southern part of their range, and foxes were suspected as the cause. In the eight years following the beginning of the experiment, rock-wallaby numbers increased by about a factor of five in the two treated sites, while one of the untreated sites remained stable and the other two declined, one of them to extinction. Subsequent experiments elsewhere in the west confirmed this finding (Saunders *et al.* 1995; Kinnear *et al.* 2002).

The results of these experiments have now been translated into management and their conclusions amplified, most significantly in the very successful Western Shield Project run by the Western Australian Department of Conservation and Land Management (see: www.calm.wa.gov.au/projects/west_shield_indep_review.html). This project takes advantage of the greater sensitivity of introduced mammals to 1080 by using broad-scale 1080-baiting as a tool to reduce fox populations over almost 3.5 million hectares in the southwest without harming native species. The initial hope was that this project would aid the recovery of a few of the rarest species in Western Australia, such as the woylie and numbat. It has done this, and under its umbrella these species have been re-established in a number of sites, but the project has also produced increases in many other mammal species as well as in reptiles and ground-dwelling birds.

Foxes are versatile predators, but their preferred prey are mammals up to about 3 kilograms (Jarman 1986); they can also handle somewhat larger animals including the juveniles of large macropods (Banks *et al.* 2000). The 'critical weight range' mammals would have been targeted by foxes, and the impact might have been increased by the fox's propensity for surplus killing – that is, killing more animals than they need and leaving many of them uneaten (Short *et al.* 2002). A recent example occurred in 1996 when a lone fox entered Garden Island after walking across the 3.7 kilometre-long causeway that connects it to the Western Australian mainland. The island is normally fox-free, and has a large population of tammar wallabies. This fox killed at least 25 wallabies in eleven days, including at least 11 in the first two days. Its killing spree ended only when it took poisoned baits specially laid for it. Other examples are given by Short *et al.* (2002). Surplus killing seems most likely in situations like the Garden Island case. A fox wandering through unfamiliar territory not occupied by a resident fox population, and among prey that have not experienced fox predation, kills continuously and indiscriminately.

Foxes can account for a large proportion of the mammal declines and extinctions, but not for all of them. For one thing, several species appear to have gone extinct before foxes arrived. This applies to all of the 'first wave' extinctions described in Chapter 10, and to the second wave of extinctions with the exception of the Tasmanian bettong. Also, there were some significant declines in the northern arid zone before the arrival of foxes, and others that extended beyond the northern limit of fox distribution. Many species of mammals vanished from the Tanami Desert before the fox reached it (Gibson 1986). The golden bandicoot *Isoodon auratus*, once widely distributed through central and northern Australia, has gone from all parts of its former range, now occupied by foxes, but also from most of the fox-free north. Its mainland distribution is now limited to a small part of the Kimberley (it also occurs on some northern islands). Some other threat was operating on these mammals, but what was it?

THE FERAL CAT

The pre-fox extinctions were of relatively small species, most of them rodents weighing under a kilogram. This is also true of most of the species that disappeared from the northern and western deserts, where foxes are scarce or absent. The desert bandicoot, pig-footed bandicoot and lesser bilby were, at around 200–300 grams, the smallest of the bandicoots, and all the rodents that disappeared in the nineteenth century were this size or smaller. This places them squarely within the size-range of mammal prey preferred by the feral cat (Dickman *et al.* 1993).

Some of the same kinds of evidence that help condemn the fox also make a strong case against the feral cat. There is a significant correlation between the presence of cats on offshore islands and the disappearance of some mammals, just as there is for foxes (Burbidge & Manly 2002). A good example is St Francis Island in the Nuyts Archipelago off South Australia. When first settled in 1880 the island was 'swarming' with rat-kangaroos (presumably woylies). They raided the kitchen gardens of the settlers, who introduced cats to control them – and in less than forty years the woylies were gone (Jones 1923–25). The abundance of feral cats predicts local extinction of conilurines rodents on mainland Australia after the effects of foxes and other factors are taken into account (Smith & Quin 1996).

Feral cats have also emerged as a serious problem for reintroduction programs. In the late 1980s and early 1990s mala (rufous hare-wallabies; see Plate 34) were reintroduced to two sites in the Tanami Desert (Gibson *et al.* 1994). At Yinapaka in the southern Tanami, 11 animals were released in August 1990. Monitoring by radio-tracking showed that they were doing well, until December 1990 when five were found dead. Warlpiri men advising on the project said that they had been killed by cats. Another three were killed in January 1991, as were more animals released later that year. These animals were killed and eaten in a characteristic fashion: their heads were chewed off and then the most accessible body flesh was eaten, especially around the chest. In many cases much of the carcass was left uneaten. Four cats were trapped, one of them a 5.1 kilogram ginger tabby found with mala hair in his stomach. Removal of these cats improved survival for a time, but eventually the population failed. At Lungkartajarra in the northern Tanami 23 mala were released in August 1989. They also did well to begin with, and the population grew to about 30 by July 1991. The first dead mala was found in October 1991, and another 14 were killed during the next four months. All were found with the marks of cat predation on them. Four cats were trapped and shot, the biggest of them weighing 4.8 kilograms. Again, this removal was followed by an increase in survival of mala, but again the population ultimately went extinct.

The experience of both programs was that cats can hunt and kill animals as large as mala, and that the killing began suddenly. Very probably this reflected the behaviour of individual large cats who learned how to hunt mala and then specialised on killing them until all were gone. Cats have contributed in whole or part to the failure of other attempted reintroductions (Dickman 1996b; Short *et al.* 1992).

Hard lessons like this have made it clear that cats can be a very serious threat to mammals up to hare-wallaby size. But for many years cats have been relegated to a minor role in the major mammal declines of the nineteenth and early twentieth centuries, because it was thought that they had been in Australia for a long time, perhaps arriving with Dutch shipwrecks on the western coast of Australia or with Indonesian fisherman as much as two centuries before European settlement. This was suspected because when European explorers first travelled through the central and northwestern deserts, they found cats. Aboriginal people had traditional names for them and claimed that the cats had always been there. This implied that the species that declined in the last two hundred years had coexisted with cats for perhaps two hundred years before.

However, Abbott (2002) comprehensively reviewed historical sources on cat distribution in Australia, and showed that cats did, in fact, arrive with British settlers and not before. They first became established around permanent settlements on the coast in the 1830s and 1840s, gradually spread inland from the southeastern and southwestern corners of the continent until about 1880, and very soon after extended their range over the rest of the continent (Figure 11.3). This means that their initial spread did correlate in time and space with the first and second waves of extinction (see Figure 10.2). It also explains why desert Aborigines said that cats had always been present, because in the northern deserts close to a human lifetime passed between the arrivals of cats and the British.

There are two other observations that seem inconsistent with a major impact of cats on native mammals. First, cats occur throughout Australia, so their geographic distribution does not fit the pattern of mammal extinction. In particular, feral cats are widespread in Tasmania, but small ground-dwelling mammals remain abundant there. If cats caused extinctions, therefore, their impact must have been limited by habitat. There is actually good evidence that this is true.

Burbidge & Manly's (2002) analysis of island extinctions revealed an interactive effect of climate and cats: cats were associated with extinctions on islands with low rainfall, but not with high rainfall. This is not surprising. Cats rely on their excellent eyesight for hunting and, as predators of medium-sized, active, ground-dwelling mammals, they are most lethal in open habitats that are found where rainfall is low (Dickman 1996b). In moist habitats with dense ground vegetation, mammals can more easily avoid cats. Abbott (2002) suggested that cats did not cause extinctions in areas with annual rainfall above 500 millimetres; Short (2004) thought the threshold might be more like 350 millimetres, and that in drier areas cats were a very significant threat. Dickman *et al.* (1993) suggested that as many as ten species of native mammals disappeared from western New South Wales as a result of predation by cats.

Second, there is the fact that cats had been a presence in the northern deserts for almost eighty years before mammals went extinct. Why would they have become a threat only recently? Again, there might be a simple explanation for this. Cats were one of the most sought-after prey of desert Aborigines, and they were easy to hunt in open country because they had little endurance in a chase and tried to escape by climbing small trees, from which they could be easily caught (Burbidge *et al.* 1988). Finlayson (1961) suggested that hunting limited the population size of the desert feral cat, commenting that 'its numbers are moderate, and as the natives hold it in high esteem gastronomically, it may possibly be checked somewhat, wherever there are active hunting populations'.

The best predictor of mammal extinctions in these regions seems to have been the decline and departure of Aboriginal populations, and the reason might be that hunting of cats by people had protected native mammals from the worst effects of this predator. We have no observations of a population explosion of cats following Aboriginal departure from places like the Tanami Desert – but then there was nobody around, black or white, to watch what happened.

Many Aboriginal people recognise that their leaving the land was linked to the disappearance of native mammals, but they interpret the relationship differently. The mammals that vanished were culturally important to traditional people, and they were central to many ceremonies intended to maintain the mammals' fertility. When people left their traditional lands they stopped performing those ceremonies. They see this failure as having caused the declines, and they blame themselves for it (Burbidge *et al.* 1988).

To summarise, there is very strong evidence of a direct role of foxes in mammal extinctions, and a less transparent but nonetheless compelling case that cats were also a significant predator. The only clear exceptions are the toolache wallaby, which was probably exterminated by habitat clearing and trophy hunting – although even for this species, Finlayson (1927) believed that fox predation was a greater problem – and the thylacine, in the extinction of which the cat's role could only have been to introduce a fatal disease. Otherwise, cats and foxes can account for practically all of the mammal extinctions of the last two hundred years.

12

Interactions: rabbits, sheep and dingoes

WHY DID FOXES and cats have such terrible effects on Australia's native mammals? These predators are among the most widespread of all mammal species, but their destructive effect in Australia seems to have been unique – unique, that is, for the continents: introduced predators have caused many extinctions on small oceanic islands (Hurles *et al.* 2003; Steadman 1995). The reason seems to be that in Australia, interactions with other environmental factors were very significant in magnifying the effects of predation on mammals in the critical weight range. The most powerful of these interactions involved introduced herbivores, especially rabbits and sheep.

THE GREY BLANKET

European rabbits *Oryctolagus cuniculus* were introduced to mainland Australia in 1860 when the wealthy grazier Thomas Austin imported wild rabbits from England and released 13 on his property, Barwon Park, near Geelong. Previous introductions of rabbits had failed, partly because native predators overwhelmed the small rabbit populations. But gamekeepers and sportsmen at Barwon Park destroyed as many of these as possible to give the rabbits a chance: in 1866, for example, 448 hawks, 23 eagles and 622 quolls were killed (Rolls 1969). There were probably more introductions at the same time, or very soon after, in other parts of southeastern Australia. Rabbits spread quickly in what effectively became a single wave that crossed the Murray River into New South Wales in 1880, reached Western Australia in 1900, and by 1910 had covered practically the whole of Australia south of the Tropic of Capricorn (Figure 11.3). Their numbers thin out north of the Tropic, and they are absent from large areas of northern Australia.

Australia's rabbits might have come from England, but they had become established there only in the late twelfth century, probably from stock that either Henry II or Richard I brought with them on returning from the Crusades (Long 2003). Their original home was Spain. The dry, open vegetation of much of southern Australia, and especially the Mediterranean climate of the southeastern and southwestern quarters, therefore suited them very well.

Before the introduction of myxomatosis in 1951, rabbit numbers in southern Australia were astronomical. The property that Eric Rolls's father bought in 1930 near Narrabri in New South Wales was typical:

> There were rabbits under the house, there were rabbits in the geraniums, rabbits ate wheat with the fowls. There was one warren on a piece of sandy soil that covered three acres. There were rabbits that one could jump on as they lay under the grass clumps in the summer. The acres of high, dead, variegated thistles – the whole property – rattled and thumped with rabbits. (Rolls 1969)

Rabbits feed selectively on soft nutritious plants and in good seasons their numbers grow until the pasture is eaten out, when they begin to dig for roots and eat the bark of trees and shrubs. In drought they eat any vegetation within reach and even climb into small trees to browse their leaves. Michael Terry, travelling through central Australia in 1931 at the end of a severe drought, saw dead rabbits caught up in shrubs at head height to a rider on camelback (Johnson & Roff, 1982).

In the early days of the rabbit invasion the combination of drought and overgrazing set hordes of desperate rabbits moving across country in such numbers that they practically covered the ground in a solid mass. At Mutooroo Station, southwest of Broken Hill, the manager tried to stop one of these plagues with a netting fence along the boundary of his property facing the rabbit-infested country, and sent two men out to patrol the fence and kill any rabbits that broke through. After a couple of days the men returned to describe what happened as the rabbit wave broke over the fence. The foremost rabbits hit the wire netting and were crushed and smothered by more rabbits coming from behind; the pile of corpses grew until it came level with the top of the fence, and the continuing stream of rabbits ran over it (Ratcliffe 1947).

Rabbit plagues turned into deserts large areas of productive well-vegetated country, much of which has never properly recovered. This must have had severe effects on the ground-dwelling mammals that relied on the same herbage as the rabbits. There are two other ways in which rabbits might have contributed to mammal declines. First, rabbits moved into the burrows of mammals such as bilbies and burrowing bettongs. Although there is little evidence that they aggressively evicted these native mammals, the sheer pressure of rabbit numbers must have pre-empted space in the burrows, and cohabitation also made it more certain that rabbits would eat out the vegetation closest to the living-places of the burrows' rightful owners. Second, in their attempts to exterminate rabbits, landholders killed many native mammals, as poisoning directed at rabbits was wildly indiscriminate. The local extinction of the eastern quoll in the Bega area coincided with the introduction during 1907–08 of rabbit poison carts, which were horse-drawn devices for the mechanised distribution of poison baits over large areas (Lunney & Leary 1988). Rabbit-sized mammals were caught in traps set for rabbits, and the burrows of native mammals like bilbies and wombats were fumigated and destroyed because they harboured rabbits.

THE SHEEP PLAGUE

Australia's sheep population grew fast in the late nineteenth century. As a result of indifferent management of flocks and overinvestment in the industry, stocking rates became extreme on marginal lands in the semi-arid zone. This caused severe overgrazing, which in the 1890s coincided with drought and rabbit plagues. The crisis was worst in western New South Wales, where the destructive effects of poor management were aggravated by the *Crown Lands Act 1884*, which broke up the large squatters' sprawling pastoral holdings and reallocated land among settlers on smaller properties (Lunney 2001). Half of the leasehold lands in western New South Wales were resumed by fiat of the Minister for Lands and made available to new settlers.

The existing pastoralists increased stock densities by moving their flocks into half the areas they had previously occupied. Investors provided loans to the new settlers to stock their runs, and for a time the profitability of established runs depended on the selling of breeding stock to found new flocks. This investment boom masked an underlying decline in profitability of the industry, while sheep numbers continued to increase. Between 1860 and 1894 the sheep flock of western New South Wales grew from under a million to more than 13 million, then collapsed to less than 4 million by 1902. It has averaged around 6 million since (Figure 12.1). That difference – 13 million to 6 million – is an index of the severity of the overgrazing in the 1890s. Add to this the effects of rabbits and drought, and it is hard to imagine that the ecosystems of western New South Wales had ever been struck a more savage blow.

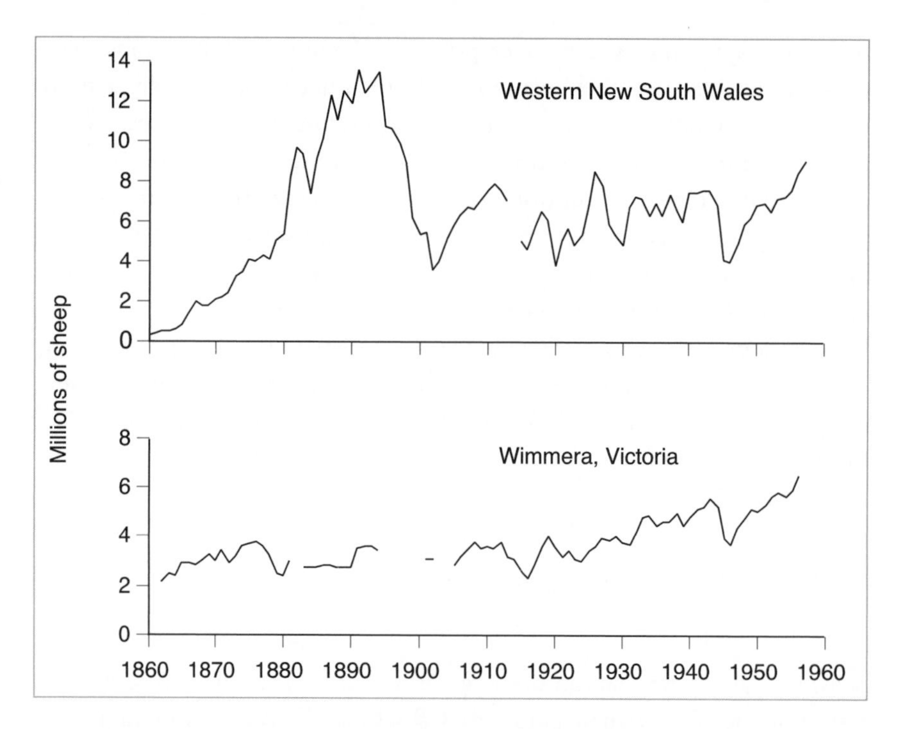

FIGURE 12.1 The number of sheep in western New South Wales and the Wimmera district of Victoria, 1860–1957 (data from Butlin 1962).

The environmental nightmare that this must have been was coolly described in the report of the Royal Commission of 1901 that inquired into the collapse of the western New South Wales wool industry: 'in the early days of settlement in the Western Division much too favourable a view was taken of the carrying capacity of the country', causing 'the destruction of almost all of the vegetation on the face of large areas of the drought-stricken country' (Lunney 2001). Lunney argued that this environmental shock was the main cause of mammal extinction in western New South Wales, where many species seem to have disappeared before foxes arrived.

INTRODUCED HERBIVORES VERSUS INTRODUCED PREDATORS

There is no doubt that rabbits and sheep overgrazed pastures and competed with herbivorous mammals, so it is common sense to regard them as having been part of the cause of the decline of those mammals. Morton (1990a) and Burbidge & Mckenzie (1989) argued that degradation of vegetation by introduced herbivores reduced the quality of the environment for native species. In particular, it eliminated critical productive patches that functioned to sustain populations of native species in droughts, and they believed this was the primary cause of extinction.

However, the extinction pattern contains a much clearer signature of the effects of introduced predators than of these introduced herbivores. There are large populations of rabbits and sheep in Tasmania, but they have caused no extinctions, nor even any obvious declines of Tasmanian mammals – Tasmania has never had a significant fox population. Ground-dwelling mammals were less likely to decline if they sheltered in burrows or rock-piles, even though this did not protect their food supplies from overgrazing – use of these refuges made mammals less vulnerable to predators. Parts of Australia that were not occupied by significant rabbit or sheep populations, for example the Gibson, Great Sandy and Tanami deserts, nevertheless suffered extinctions (Johnson, *et al.* 1989) – these areas had feral cats (as well as scattered fox populations). Carnivorous marsupials declined along with the native herbivorous mammals, even though they would not have competed directly with rabbits and should have benefited from the availability of young rabbits, at least, as prey – these species were also eaten by foxes. Large kangaroos also compete with sheep and rabbits for food but they have increased in abundance – predation on these species by cats and foxes is not significant.

Even in regions that suffered rabbit plagues, some areas were less badly affected than others. Rabbits prefer deep sandy soils and grasslands. They avoid black or clayey soils, heaths and densely wooded areas, and these habitats should have provided refuge areas to protect at least some native mammals from the worst effects of rabbits. There is historical evidence that native species did in fact survive rabbit plagues while remaining reasonably widespread and abundant. Bilbies remained common in the southwest of Western Australia for twenty years or more after rabbits arrived (Abbott, 2001). The chuditch survived in central Australia alongside rabbits through the intense droughts of 1925–30 and 1940–41, only to vanish in relatively good seasons later on (Johnson *et al.* 1982). Rabbits and burrowing bettongs coexisted in central Australia before foxes arrived and the bettongs disappeared (Finlayson, 1958). On the slopes and tablelands of New South Wales rat-kangaroos remained abundant as the sheep flock grew, and they survived the combination of rabbit plague, overgrazing by sheep and severe drought in 1902. But as foxes invaded, rat-kangaroos declined to rarity or extinction (Short 1998). Most of the examples that Lunney (2001) gives of mammals declining before the arrival of foxes are relatively small-bodied species that may well have been eliminated by cat predation. This could also apply to the Flinders Ranges, where Tunbridge (1991) argued that many mammals disappeared before foxes arrived.

Sheep were and are concentrated in the semi-arid and temperate southeast and southwest, but mammal extinctions extended over a much larger area of central Australia. Sheep came too late to explain the first wave and much of the second wave of extinctions described in Chapter 10, and they were in the wrong place to explain the fourth wave, which affected desert mammals. The massive overgrazing by sheep in western New South Wales must have had awful effects, but there were mammal declines of similar magnitude and at the same time in places where sheep numbers did not go through such an extreme cycle of boom and bust – in the Wimmera region of Victoria, for example, adjacent to western New South Wales and climatically similar to it (Figure 12.1). Fisher *et al.* (2003) found a positive correlation between the extent to which the original geographic ranges of marsupial species overlapped with sheep and the severity of range decline. This could be taken as evidence of a direct impact of sheep on those mammals, but equally it could implicate something associated with sheep, like extremely high densities of foxes.

HYPERPREDATION

While the direct effects of rabbits and sheep on native mammals might have been small compared with those of introduced predators, it is clear that rabbits and sheep had large indirect effects. They first paved the way for the spread of the fox, then rabbits maintained foxes at extremely high numbers. Rabbits and foxes were introduced into southeastern Australia on large sheep runs, and both species became most abundant in sheep country in the southeast and southwest of the continent. Land that was good for sheep grazing was also good for rabbits – it was rather dry, with generally open vegetation, light soils and a dense pasture layer. Sheep improved this habitat for rabbits by heavy grazing that created ideal short pastures, and the new landholders further obliged the rabbit by providing water and clearing some of the timber. The combination of suitable climate, plenty of rabbits, and an open but varied vegetation providing good hunting grounds with patches of cover, was perfect for foxes, and they spread most quickly where they found both sheep and rabbits (Jarman 1986). So to some extent rabbits, sheep and foxes were a package: rabbits followed sheep, and foxes followed rabbits (although rabbits and foxes ultimately spread well beyond the range of sheep).

The current distributions of the rabbit and fox over mainland Australia match one another very closely. Where rabbits are abundant, so are foxes; as rabbits thin out in northern Australia, so do foxes; and it is probably the absence of rabbits that explains why foxes have not occupied the whole of the north. The spread of the fox followed that of the rabbit with a lag of about ten to twenty years, and foxes spread fastest where rabbits were abundant (Jarman 1986). Foxes did manage to reach some areas beyond the range limits of the rabbit, such as the Kimberley region of northwestern Australia, but here their occupation is tenuous. Fox numbers in a region are still, to a very large extent, a function of rabbit numbers. As rabbit populations increase in good seasons, so, after a short time lag, do those of foxes. If rabbit numbers collapse as a result of drought or an outbreak of disease, foxes also decline, again after a short time lag (Williams *et al.* 1995).

Rabbits are the preferred prey of foxes in Australia, and when rabbits are abundant they contribute the bulk of the fox's diet. But foxes are versatile predators. While they might be largely sustained by rabbits they readily take other prey, especially other mammals of roughly rabbit-size. When mammal prey become rare, foxes turn to carrion, invertebrates, fruit, reptiles, frogs, bird and reptile eggs, and so on. Switching to alternative foods allows them to survive crashes in the populations of their preferred prey, albeit at reduced density.

The establishment of the fox–rabbit nexus in Australia therefore created two problems for native mammals. First, the enormous abundance of rabbits allowed fox populations to reach very high densities, so that the ratio of foxes to native mammals was held at a high level. Second, a crash in the rabbit population would be followed by a transitional period when a large population of foxes suddenly found themselves short of food, and predation pressure on alternative (native) prey increased. For these reasons the presence of rabbits magnified the impact of foxes on prey other than rabbits, a phenomenon referred to by Smith & Quin (1996) as 'hyperpredation'.

The effects of hyperpredation have been explored mathematically by Courchamp *et al.* (2000), who showed that where an introduced predator is maintained at high density by an introduced prey species, extinction of a native prey species is very likely if any of the following conditions apply: the introduced prey has a higher population density than the native species; the introduced prey has a higher rate of population growth than the native species; or the native species is easier to catch than the introduced species. In each case, the introduced prey species supports high predator density, but is not as strongly affected by predation as is the native prey.

The hyperpredation idea makes good sense of some otherwise puzzling cases of extinction, such as that of the Macquarie Island parakeet *Cyanoramphus novaezelandiae erythrotis*. This extraordinary parrot lived year-round on subantarctic Macquarie Island, spending most of its time near the seashore. There it fed on invertebrates gleaned from heaps of seaweed, and nested in and under tussocks. It would have been vulnerable to introduced predators, but its population was unaffected when cats were introduced to the island around 1820. In 1880 rabbits were introduced and quickly became abundant. Within ten years the parakeet was extinct. Did rabbits kill the parakeets? Or did they eat out its food supply? What seems to have happened is that rabbits provided a stable winter food supply for the cat population, which increased manyfold. Predation on parakeets by cats went up accordingly, and the parakeet was hunted to extinction (Taylor 1979).

The interaction between foxes, rabbits and mammals in Australia fulfils the requirements for hyperpredatory extinction. Rabbits reached higher densities than native mammals, and the abundance of rabbits subsidises large fox populations. Hyperpredation might have involved cats as well as foxes. Cats prey heavily on young rabbits, so their abundance would also have been boosted by the arrival of rabbits (house mice might have provided another superabundant introduced prey for cats, Smith & Quin 1996). However, the history of predator–prey interactions involving cats is more complicated than for foxes. Cats arrived before rabbits, and initially their abundance was moderate because they did not have the benefit of a superabundant prey species; also, they were probably suppressed by people in open habitats. When rabbits arrived the cat population might have increased, but perhaps not significantly where cats were still hunted by Aborigines. Finally, release from Aboriginal predation triggered population growth, especially where cats had plenty of rabbits to eat. The situation was made worse because Europeans released cats in large numbers in the hope that they would control rabbits. There are reports from the 1880s of cats being shipped by rail in batches of hundreds and let go in remote regions to supplement wild cat populations. According to Rolls (1969):

> The cats turned out for rabbit-eating were not simply shaken from a bag near a burrow and left. Portable pens were erected in the bush and the old cats were fed there until they had kittens. The pen was then opened and the feeding gradually reduced so that they would accustom themselves to bush life.

Hyperpredation was being given a helping hand.

Native mammals may have been easier prey for cats and foxes than were rabbits, at least initially, because of naivety. Australian mammals had coped with mammalian predators before foxes arrived and they had appropriate fear and avoidance behaviours to minimise predation risk (Blumstein & Daniel 2003). They were not hopelessly naive in the way that species on small predator-free islands might have been. But at first they may not have responded to foxes as the formidable predators they are, and they might have needed some time to calibrate their anti-predator

Plate 35 Northern bettong *Bettongia tropica*, one of the rat-kangaroos that specialises on a diet of subterranean fungi ('truffles'). (photograph by Karl Vernes)

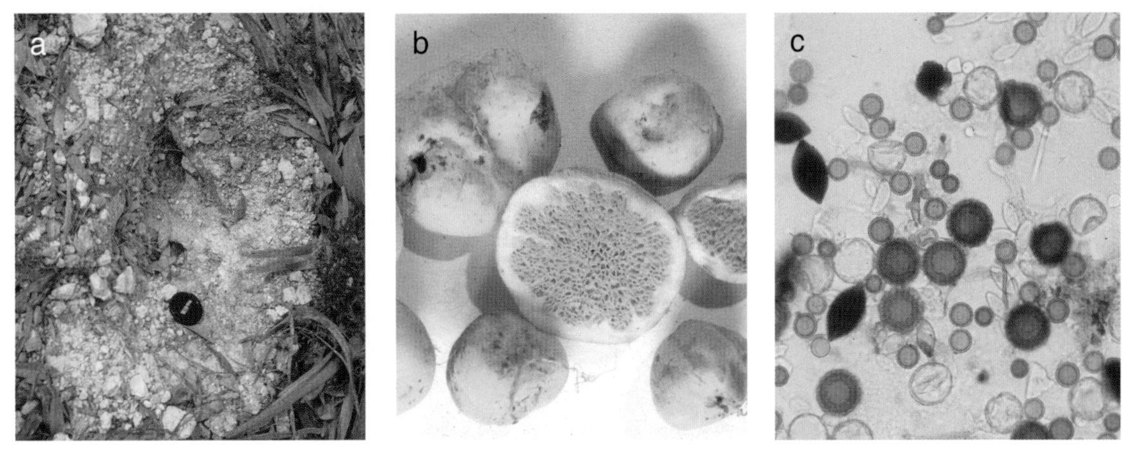

Plate 36 Truffle-eating by bettongs: (a) digging made by a Tasmanian bettong in search of truffles; (b) examples of truffles – these are *Beatonia* sp. from north Queensland (photograph by Karl Vernes); (c) diverse fungal spores extracted from faeces of the Tasmanian bettong.

PLATE 37 The greater bilby *Macrotis lagotis* disappeared from the southern and eastern parts of its original range and now has a patchy distribution in the deserts of central and northwestern Australia. It has been reintroduced into predator-free reserves in South Australia and western New South Wales (photograph by Rick Stevens/Fairfaxphotos.com).

PLATE 38 The bridled nailtail wallaby *Onychogalea fraenata* was reduced to one small population in central Queensland. It remains rare, but has been reintroduced to several localities in Queensland and western New South Wales including the Australian Wildlife Conservancy's Scotia reserve. (photograph courtesy of the Australian Wildlife Conservancy)

behaviour to include this new enemy. There is evidence that such calibration is possible. In Tasmania eastern quolls do not react to foxes as predators, but they do react appropriately to cats – foxes are still a novel predator in Tasmania but cats have been there for well over a hundred years (Jones *et al.* 2004).

Further, rabbits as well as foxes and cats have higher population growth rates than native mammals. A burrowing bettong, for example, produces no more than three offspring in a year, but a vixen can have up to ten cubs, and a cat ten or more kittens, in the same time. A growing fox population can therefore produce new foxes much more quickly than a bettong population can replace animals eaten by foxes. Rabbit populations, with females producing twenty or more kittens in a year, are demographically more resilient to the high intensities of predation that cats and foxes are able to impose. In a predator–prey system consisting of foxes, rabbits and bettongs, the bettong is predisposed to suffer unsustainable rates of predation. This would have been especially true under the very unstable conditions that followed the introduction of rabbits and foxes, when populations of both those exotics increased at their fastest possible rates.

Reproductive rates are higher in smaller species, and this made the smallest marsupials and rodents more resilient to predation than the species in the critical weight range. Smith and Quin (1996) found that among conilurine rodents, declines were greatest among species with the lowest reproductive rates. This relationship helps to explain why there have been no significant declines among native rodents other than conilurines: the non-conilurine rodents have higher reproductive rates than the conilurines (Yom-Tov 1985). There may be a similar relationship between severity of decline and reproductive rate among marsupials (Cardillo 2003), but demonstrating it is difficult because of lack of data on reproduction in extinct species (Fisher *et al.* 2003).

Despite suffering these disadvantages, native mammals can coexist with exotic predators under some conditions. There is historical evidence for this. The dingo caused few or no extinctions when it first arrived in Australia; many native mammals on the mainland coexisted with the cat for nearly a century and still do in some habitats; and in Tasmania feral dogs were common enough to be a serious pest to the sheep industry in the nineteenth century but caused no extinctions of native species.

In an important study, Sinclair *et al.* (1998) analysed demographic data on populations of critical-weight-range mammals reintroduced into environments containing exotic predators. By fitting these data to classical predator–prey models, they demonstrated the existence of thresholds of population density above which the natives should be able to absorb the effects of exotic predators, but below which they decline to extinction. The reintroduction programs they analysed tried to establish populations below these thresholds, and were doomed to failure – typically, animals were released in groups of less than thirty individuals. Sinclair *et al.* concluded that releasing hundreds rather than tens of animals could have established viable populations. Management aimed at reducing predation rates by suppressing predator density or by making predation less efficient (for example, by encouraging dense ground vegetation to provide cover) would have the same effect.

WHY AUSTRALIA NEEDS MORE DINGOES

We have now looked at all the factors that could plausibly have caused extinctions, and isolated predation by foxes and cats as the most general and destructive. I think it is fair to say that had foxes and cats never been brought to Australia, none of the recent mammal extinctions (except those of the thylacine and, possibly, the toolache wallaby) would have happened. Some species might have become rare under pressure from rabbits and livestock, and we might now be wondering what should be done to prevent further declines of the pig-footed bandicoot, the 'wide-awake', or oolacunta, but these animals would still survive in reasonable numbers.

The hyperabundance of rabbits goes part of the way to explaining the extreme impact of introduced predators, because it allowed fox and cat populations to reach extraordinary densities. The other part of the explanation is that when they were released in Australia, foxes faced no significant predators of their own. The same was true of cats when Aborigines stopped hunting them, except that foxes became an important predator of cats (see below). The one species that might have been an effective biological control of both foxes and cats had been, to a large extent, put out of action.

Just how abundant and ubiquitous dingoes were when Europeans first settled in Australia is unclear. Probably they were initially uncommon but their numbers then increased in settled districts as a result of the provision of water for livestock, the removal of Aboriginal predation (see Chapter 9), growth of the kangaroo population and, of course, the availability of mutton (Corbett 1995). Dingoes were seen by the settlers as a menace to livestock and were persecuted accordingly. This persecution was especially intense around sheep, because dingoes find sheep easy to kill. In New South Wales the government bounty on dingo scalps was as high as four pounds, double the maximum amount offered for any other species (Glen & Short 2000). Persecution of dingoes was quite general, and by the end of the nineteenth century they appear to have been rare in many places. In 1922 Charles Hoy collected a dingo near Ravenshoe in north Queensland. Hoy, an expert mammalogist who had at that time spent more than two years collecting mammals throughout Australia, wrote, 'Strange to say it is only the fourth dingo I have seen and the first I had a shot at' (Short & Calaby 2001).

Suppression of dingoes meant that by the time foxes began their invasion of Australia the only large mammalian predator on the continent had been neutralised as an ecological force. Other native predators such as quolls and eagles do not trouble foxes, so there was nothing apart from drought and fluctuations in the supply of rabbits to check their increase. Destruction of dingoes in New South Wales peaked between 1891 and 1900, just as foxes were becoming common (Glen *et al.* 2000); by 1910 dingoes had been exterminated in some districts, while foxes were driving rat-kangaroos extinct. Dingoes have been eradicated from most of New South Wales, apart from the eastern forested regions where sheep grazing is not significant, from most of Victoria, and from the southwest of Western Australia (Fleming *et al.* 2001).

Foxes spread most quickly in country without dingoes, such as the sheep runs of the southeast, and more slowly in cattle country and beyond the settled districts where dingoes remained common (Jarman 1986). Foxes continue to be most abundant where dingoes are rare, and vice versa (Glen & Dickman 2005). This is true at the continental scale (Smith *et al.* 1996), and also within regions. Figure 12.2 provides an example of the inverse relationship between dingo and fox abundance for southeastern New South Wales (Nadgee Nature Reserve and Kosciuszko National Park; Newsome, 2001). This relationship is triangular – when dingoes are scarce, foxes may be abundant but are not always so, presumably because factors other than the presence of dingoes also affect their numbers. However, when dingoes are abundant, foxes are invariably rare. This suggests that dingoes set a limit on the maximum abundance of foxes. The boundary that defines the dingo/fox tradeoff in this case has a slope of close to 45 degrees, implying that, very approximately, any increase in the dingo population produces an equivalent decline in the fox population.

Newsome *et al.* (2001) compared communities of large vertebrates on either side of the dingo barrier fence between New South Wales and South Australia. On the New South Wales side, where dingoes have been eradicated, fox density was up to twenty times higher than on the South Australian side, which still has dingoes. A study of the distribution of

FIGURE 12.2 The negative relationship between abundance (indicated by number of tracks per 100 metres of raked roadway) of foxes and dingoes in Kosciuszko National Park and Nadgee Nature Reserve, New South Wales (data from Newsome 2001).
Notes: There is a negative linear relationship between the variables but it is weak and non-significant ($F_{1,13} = 13.36$, $p = .02$). I showed the significance of the triangular relationship by sorting estimates of dingo abundance into classes 4 abundance units wide, and regressing the maximum fox abundance in each class against the mid-point dingo abundance for that class. This regression was significant ($F_{1,4} = 13.36$, $p = .02$), and the regression equation was used to find the location of the upper bound of the triangle shown in the figure.

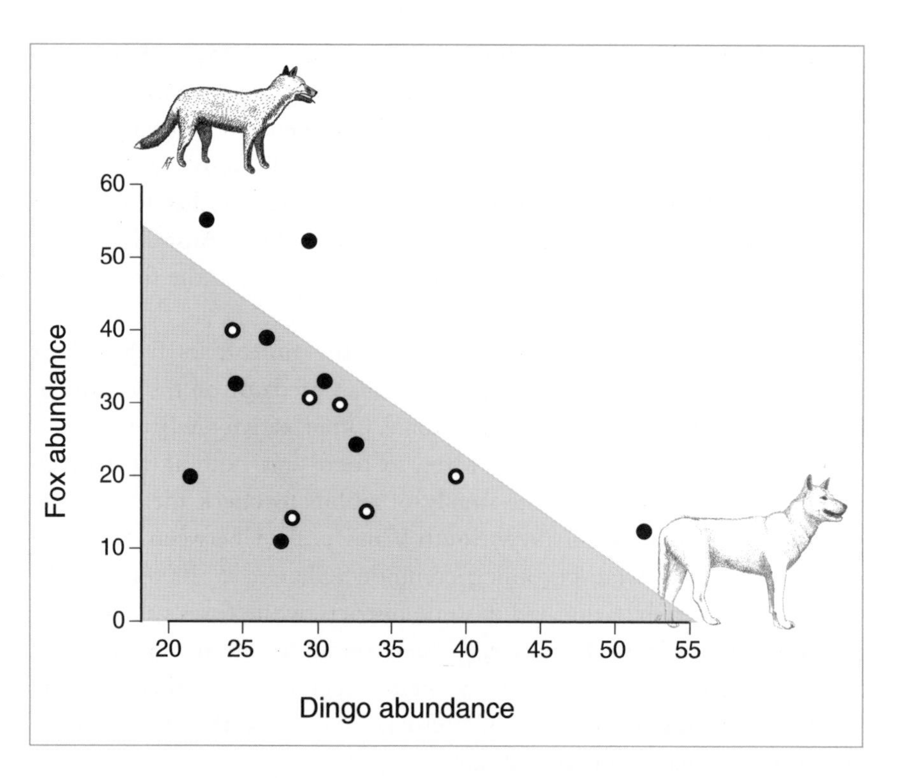

dingoes and foxes in the Blue Mountains west of Sydney recorded tracks around bait stations, and found that foxes were less likely to use bait stations that had been visited by dingoes (Mitchell & Banks 2005). In this case there was no relationship between dingo and fox abundance at larger scales, but as both species were being heavily suppressed by poison baiting, it is possible that this pre-empted the effects of interactions between the two species on their abundance.

Dingoes probably suppress foxes in a number of ways. Although they tend to prefer large mammals and foxes take smaller prey, there is broad overlap in the diets of the two species, so the presence of dingoes could reduce availability of prey for foxes (Glen *et al.*, 2005; Mitchell & Banks 2005). In central Australia dingoes excluded foxes and cats from carrion and waterholes during a drought when food and water were scarce, and the two smaller predators disappeared locally as a result (Corbett 1995). Dingoes also kill foxes. This is partly for food, and some studies show that foxes are eaten regularly enough to be regarded as a significant prey item for dingoes. But it also seems that dingoes will go out of their way to kill foxes whether they eat them or not. Just as dogs hate cats, dingoes evidently hate foxes.

The ecological sense behind this is that by killing foxes dingoes eliminate a potential competitor. 'Competitive killing' of this kind is common among large predators. It has been reviewed by Palomares &

Caro (1999), who showed that of all the world's carnivores the grey wolf and the red fox have the greatest propensity for malicious killing of other predators (and remember that the dingo is a wolf). Wolves in other countries kill adult foxes without eating them, and they dig fox cubs out of their dens to murder them (Jarman 1986). Foxes therefore fear dingoes and avoid them, and the effect of a dingo population on foxes is greater than if foxes simply fell prey to dingoes from time to time.

There appear to be similar interactions between dingoes and cats, and foxes and cats (Glen *et al.* 2005): cats are rare where either dingoes or foxes are abundant. When foxes are reduced, cats may increase. At Herrison Prong, a semi-arid site without dingoes on the Western Australian coast, fox control was followed by a three-fold increase in the abundance of cats (Risbey *et al.* 2000).

The evidence that dingoes suppress fox numbers is largely circumstantial. Nobody has yet conducted a controlled experiment to test whether reducing dingo populations allows foxes to increase, or if allowing dingoes to increase causes foxes to decline. The need for such an experiment is urgent, but there is enough evidence to justify the following statements as solid working hypotheses. First, the reason that foxes were able to reach such high densities in Australia and cause so much destruction was that dingoes were rare. Second, continued control and eradication of dingoes allows foxes to remain at high numbers and to suppress native mammals over most of southern Australia.

In the four thousand years or so that dingoes have been in Australia they have taken over from the thylacine as the largest predator on the continent, and in particular as the main hunter of kangaroos. There is strong evidence that predation by dingoes reduces population density of the larger kangaroos, and it is possible that it acts in a regulatory fashion to stabilise kangaroo numbers (Fleming *et al.* 2001). Dingoes have also taken over from devils as the only large mammalian scavenger on mainland Australia. The dingo is clearly an ecologically significant species that interacts strongly and stably with other native mammal species. But, it has not been implicated in any extinctions of native mammals other than the thylacine and devil – and, as I argued in Chapter 9, the evidence linking the dingo to these extinctions is weak.

We know of no other mammal extinctions connected to the arrival of the dingo. It is possible that dingo predation added to the pressure that caused so many critical-weight-range mammals to go extinct in the recent past, but this seems unlikely. Species of rodents that live in parts of Australia where dingoes remain common have experienced less decline than species from areas where dingoes are rare or absent (Smith *et al.*

1996). The geographic pattern of mammal extinction is almost the inverse of the distribution of high-density dingo populations (compare Figures 11.3 and 10.2).

Why, if dingoes are such significant predators, have they not themselves caused extinctions of native prey? There are probably several reasons. Dingoes live in large and stable territories, from 20 to several hundred square kilometres in area depending on the productivity of the habitat. Each territory is occupied and defended by a single family group, usually with only one breeding female, so large territories mean low population densities (Fleming *et al.* 2001). When kangaroos or, more recently, rabbits and sheep are abundant, dingoes prey heavily on them rather than on smaller native species. Dingoes have slower rates of increase than foxes and their populations tend to be more stable. Foxes have a similar social organisation but they occupy much smaller territories of five square kilometres or less, and when prey are very abundant the territorial system may break down entirely, allowing groups to use overlapping ranges at very high abundance (Saunders *et al.* 1995). Fox populations are therefore more volatile than those of dingos and, because foxes prey more intensively on middle-sized species, they can have very large impacts on such species.

This interaction of dingoes and foxes is an example of a phenomenon called 'mesopredator release'. Surprisingly, prey species may be safer from predation when they are exposed to more predator species, because interactions among various predator species minimise the impact of any one of them (Finke & Denno 2004). In particular, large and aggressive top predators, such as wolves and big cats, suppress populations of intermediate-sized predators ('mesopredators'). These mesopredators are typically versatile generalists with high capacity for population growth. Removal of top predators allows them to increase and exposes their prey to high predation rates.

The effects of mesopredator release can be seen in fragmented habitats. In Californian woodlands coyotes *Canis latrans* suppress feral cats but, in habitat fragments that are too small to sustain coyote populations, cats increase, predation rates on small vertebrates go up, and songbirds are killed out (Crooks & Soule 1999). Essentially the same process has been repeated many times on islands when typical mesopredators such as cats and rats have been released into environments lacking larger predators capable of suppressing them (Courchamp *et al.* 1999). The experimental reduction of foxes at Herrison Prong demonstrated an effect of release of a mesopredator (in this case, the feral cat) on populations of small mammals: after foxes were controlled, small mammals in the preferred prey-size range of cats declined by 80 per cent (Risbey *et al.* 2000).

The history of Australian mammals over the last two hundred years can be viewed as a vast experiment on the effects of mesopredators release. Removal of top predators – dingoes and Aboriginal people – exposed smaller prey to high intensities of predation from some combination of foxes and cats. The effects of this were made worse by the extraordinary abundance of introduced prey, which fuelled extreme population growth of those predators.

It is possible that changes to habitat also played a role in magnifying the impacts of predation on critical-weight-range mammals. Although changed fire regimes probably did not harm native mammals directly (see Chapter 11), increased fire removed vegetation cover and left native mammals more exposed to cats and foxes. The difference in the fate of bettongs after fire on the mainland and in Tasmania provides a good example of this effect. In the southwest of Western Australia woylies (brushtail bettongs) live in habitats that are frequently burnt, but they have wonderful fire-survival skills that allow them to remain within their normal feeding and nesting areas as fire passes over them. Even in very hot fires they flee only short distances, then double back through breaks in the fire front, or even dash straight through the flames if necessary, to stay close to home (Christensen 1980). They find plenty of food in recently burnt areas (in the form of truffles, which become either more abundant or more accessible after fire), but many are killed by predators. The responses of Tasmanian bettongs to fire are very similar, except that in the absence of foxes their survival after fire is high (Johnson 1995). In the Simpson Desert cats and foxes periodically invade severely burnt habitat and cause local extinctions of small mammals after widespread fires (Dickman 1996c), presumably because with the removal of cover by fire small mammals are easier to hunt. Vegetation clearing might have worked against native mammals in the same way.

This perspective might help to explain some worrying declines of mammals that have recently been detected in several parts of the Northern Territory and in the Kimberley region of Western Australia (Price *et al.* 2005). So far, these declines seem to have been patchy but have affected a large area in total, including significant protected areas such as Kakadu National Park (Woinarski *et al.* 2001). The species involved have been small and medium-sized ground-dwelling mammals, some of which have recently become rare on the northern mainland and remain common only on islands. It is widely suspected that vegetation changes due to livestock grazing or increased fire might be driving these declines – except that they extend beyond pastoral lands, and that research has so far uncovered no simple relationship between burning patterns and mammal decline (Price *et al.* 2005).

There are no foxes in the north, but there are cats, and we know that under some conditions cats can be very destructive predators. One possible explanation of what is happening is that fire or grazing has recently increased the exposure of ground-dwelling mammals to predators, and that at least in some places this has allowed cats to become a significant threat. The evidence appears strongest for the northern quoll *Dasyurus hallucatus*. This species began to decline first in open savannas where cattle grazing and frequent fire had removed understorey vegetation, and its distribution has contracted to isolated refuge areas with rocky landscapes and dense vegetation cover (Jones *et al.* 2003; Braithwaite & Griffiths 1994). A demographic study of one declining population found a high mortality rate due to predators (feral cats and dingoes) and showed that females with access to rocky habitat survived best (Oakwood 2000).

The pattern of these changes to mammal communities in northern Australia is alarmingly similar to the earlier losses of medium-sized ground-dwelling mammals from most of the mainland of southern Australia. Discovering their cause is an urgent research problem. If it is left unsolved, we might find that we are watching the unfolding of a fifth wave of mammal extinction to add to the four described in Chapter 10.

A COMPLETE MODEL AND ITS IMPLICATIONS

The discussion so far is summarised in Figure 12.3, which offers a model of the interactions that led to the declines and extinction of critical-weight-range mammals in Australia. There are two major parts to the model: the central and most significant interaction of introduced predators bearing down on native prey; and a set of other factors that intensified this core interaction. The most important of these other factors were those that raised population densities of cats and foxes. Rabbits and other introduced prey species provided a large prey base for them; suppression of dingoes gave them a free hand; the departure of Aboriginal people removed a check on cat numbers; and sheep (and their owners) contributed by improving habitat for rabbits. Fire and livestock grazing reduced the density of ground-vegetation, making it more difficult for small and medium-sized mammals to avoid predators.

These co-factors explain why cats and foxes had such large effects. The very distinctive pattern of extinction can then explained by two major features of prey species:

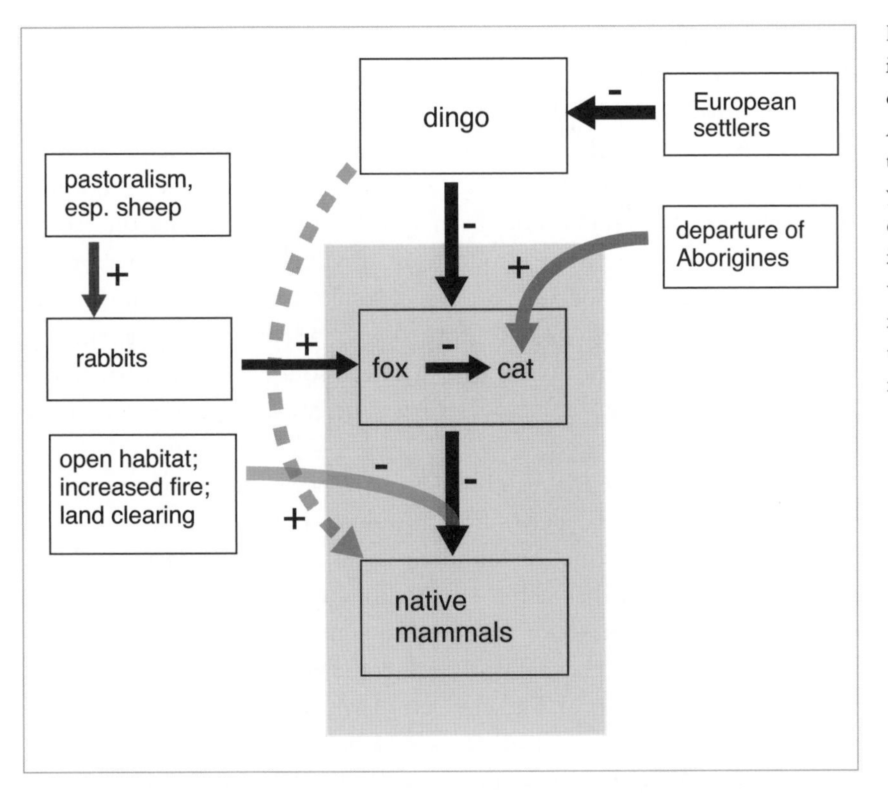

FIGURE 12.3 Summary of interactions that led to declines and extinctions of Australian mammals over the last two hundred years. Arrows show the direction of effects, and indicate whether they were positive (+, led to an increase in abundance of the target organisms) or negative (-).

Body size Species small enough to be the prey of foxes and cats suffered most, but the very smallest animals were less affected because their higher population growth rates made them more resilient to predation.

Distribution and habitat Species that were most exposed to cats and foxes and suffered the most severe declines were those whose geographic ranges overlapped with foxes and that preferred open habitats where foxes and cats hunt most efficiently, lived on the ground, and did not use rock-piles as refuges.

There are other pressures on native mammals in Australia, some of which are growing in significance. For example, goats compete with rock-wallabies, cane toads are currently threatening quoll populations across the north, and habitat fragmentation erodes the population viability of forest-dwelling mammals in the south. Climate change is looming as a very large threat. New diseases are emerging in isolated mammal populations, most frighteningly in Tasmanian devils. Nonetheless, cats and foxes take first place as destroyers of mammal populations, and this is as true now as for the last 150 years. Neutralising these predators is an essential condition for the protection and recovery of those critical-weight-range mammals that have managed to survive thus far, and, unless this is done, tackling the other threats will probably be futile. So far, success in the suppression of cats and foxes for mammal conservation has been achieved on mainland Australia in two ways.

First, broad-scale 1080 baiting has been used in the Western Shield program (see page 202) to reduce fox populations across large areas of southwestern Australia. Populations of native mammals have recovered very well since the program began in 1996, and in many cases the recoveries have been helped by reintroductions, of which there have been more than 60, involving 16 species. As a direct result, three rare mammals have had their official threat classifications downgraded: the woylie, tammar wallaby, and quenda (the western form of the southern brown bandicoot *Isoodon obesulus*). Many other species have increased in abundance in the 3.5 million hectares treated.

Second, some large mainland refuges have been created by eradicating cats, foxes and other pest species from nature reserves that are fenced off to prevent reinvasion from surrounding areas. The Australian Wildlife Conservancy (AWC) has achieved this in several large nature reserves around Australia, the largest of which is Scotia in western New South Wales where foxes and cats have been excluded from 12 000 hectares (see www.australianwildlife.org). These reserves have been successfully restocked with mammal species that had previously been locally exterminated by predators. The AWC has so far undertaken reintroductions of 13 mammal species to its reserves (Schmitz 2005), and the work at Scotia should see the re-establishment in New South Wales of mammals such as the bilby (Plate 37) and bridled nailtail wallaby (Plate 38), currently listed as extinct in that state. Exclusion of pest animals and reintroduction of native species has been similarly successful in the 6000 hectare Arid Recovery reserve at Roxby Downs in South Australia (www.aridrecovery.org.au).

Both these approaches have worked, but there are limits to how widely it will be possible to apply them. Western Shield depends on the evolutionary legacy of high tolerance of 1080 by mammal species in southwestern Australia (within the range of the various species of 1080-bearing poison bushes), which makes possible its liberal use as a tool for conservation. Suppression of predators by other means requires huge effort and expense, and fencing them out becomes increasingly difficult as the size of the area to be protected increases. We will need other solutions to put mammals back into the vast areas that surround the refuges created by 1080 baiting in the southwest and by predator exclusion from isolated reserves elsewhere. Biological control of predators could achieve this. More than ten years' work has gone into the development of a biotechnology-based biological control strategy to reduce fox populations, most of it by the Invasive Animals Cooperative Research Centre and its predecessor organisations (www.pestanimal.crc.org.au). This may or may not produce a practical control agent, but even if it does the threat from cats would remain, and probably grow worse as foxes declined.

There is another way to think about the problem. European settlement of Australia changed the continent's landscapes in many ways, but a remarkably consistent feature of the changes was that they created environments that were ideal for cats and foxes, and therefore dangerous to native mammals that were of the right size and lived in the right places to serve as prey for those two predators. Could we redesign Australian landscapes to tip the balance back in favour of native mammals without actually eradicating cats and foxes? The model illustrated in Figure 12.3 identifies several ways in which this could be done.

Perhaps the most important, and the most practical, of these is to increase the abundance of dingoes. The dingo may well be the most effective biological control agent for both foxes and cats that we will ever have, but management of dingoes consists largely of continuing attempts to reduce or eliminate their populations. If this attitude changed, and if dingoes were encouraged at moderate densities in southern Australia, we might reap huge benefits for conservation. The difficulties in doing this vary regionally, and they depend mainly on whether the dominant livestock are cattle or sheep. Dingoes are only a moderate threat to cattle. Some cattle graziers persecute them, but others tolerate them on the grounds either that persecuting them is not worth the effort, or that a healthy population of dingoes reduces kangaroo numbers and thereby provides an indirect benefit to cattle production. There is even circumstantial evidence, for example from Newsome *et al.*'s (2001) ecological comparisons of the two sides of the dingo exclusion fence on the border of New South Wales and South Australia, that dingoes help control other pests such as feral goats and pigs.

It is not at all obvious that the benefits to cattle graziers of controlling dingoes (mainly, increased survival of calves) outweigh the costs of doing so (the time and expense expended on control, and the loss of the benefits of dingo predation on kangaroos and other pest species). A formal cost-benefit analysis of this problem would be most interesting, and might well show that dingo control is unwise when judged simply as an economic input to beef cattle production. There is very little doubt that if the conservation values of dingoes were also put onto the scales, dingo persecution would be seen as a very poor strategy.

Matters stand differently in sheep country. Dingoes are a serious threat to sheep, so much so that uncontrolled dingo populations can make sheep-raising unviable. Some level of dingo control will probably always be essential for the sheep industry, but even here it is not clear that dingoes should be eliminated. Foxes are significant predators of newborn lambs, at least in some circumstances (Saunders *et al.* 1995). In a cost-benefit calculation of dingo control in sheep-grazing lands, the damage done by dingoes would be traded off against their effect in reducing fox populations. Figure 12.2 suggests that the relationship of the abundance of the two species might be approximately one-to-one. This does not mean that the trade-off between them is neutral, because dingoes kill more sheep than foxes do. But it could mean that even for sheep graziers, dingoes confer some benefits as well as exacting severe costs; and as noted above, the benefits might include control of kangaroos and so on as well as suppression of foxes. The costs to sheep production of retaining even a low density of dingoes would probably outweigh the benefits. But might it be possible to calculate the true costs to sheep farmers of accepting a certain level of dingo population, and to pay them a subsidy as compensation to secure the conservation benefits of dingoes?

The destructive power of foxes and cats would be further undermined if there were fewer rabbits in Australia. In fact, the recent spread of rabbit haemorrhagic disease has reduced the rabbit population by more than 80 per cent through large areas of the southern arid and semi-arid regions (Cooke & Fenner 2002). One of the initial concerns over introducing this disease (or any other control agent) into the Australian rabbit population was that with the sudden loss of their major source of food, cats and foxes would increase their impact on native species. This has not happened. In some areas the decline of rabbits had little effect on introduced predators and native species (Edwards *et al.* 2002); in others the decline of rabbits was followed by declines or even disappearance of cats and foxes, and a consequent reduction in total predation pressure on native prey (Read & Bowen 2001; Holden & Mutze 2002).

Finally, the exposure of ground-dwelling mammals to predators could be reduced by restoring to the vegetation some of its original structure and diversity. The combined effects of vegetation clearing, of heavy grazing by cattle, sheep and rabbits, and of frequent fire, create a very open and simple shrub and herb layer. This allows maximum freedom of movement for cats and foxes, and provides minimum concealment for their prey. Management of fire and grazing to allow some recovery of more complex and dense ground vegetation would reduce the impacts of predators, even if those predators remained moderately abundant. Where this is in conflict with economic land uses – such as sheep grazing, with its requirement for short, uniform (and biologically sterile) pastures – the creation of predator-proofing vegetation could be planned as a patchwork of refuges distributed through pastoral lands, each large enough to benefit small viable populations of a few mammal species.

With the combination of these three changes – more dingoes, fewer rabbits and a more dense and complex ground-vegetation layer – reintroductions of native mammals might succeed even without pre-treatment to remove introduced predators. All three of these improvements could be achieved by modifying rather than replacing existing land uses over large areas, and they could be made consistent with cattle and (with a bit more difficulty) sheep production. Together, they are a prescription for healthy environments for Australian mammals. Putting this prescription into practice should be the goal of land management for mammal conservation in Australia.

13

Conclusions

WHAT DID THE three major waves of extinction have in common, and how might they have been connected to one another? The extinctions involved a very diverse collection of species, from the giant herbivores and spectacular predators of the late Pleistocene, to medium-sized carnivores in the Holocene, and much smaller bandicoots, rat-kangaroos, rodents and so on in the last 200 years. And they unfolded across a wide range of environments, over vast distances in space, and in the very different climate regimes of the last glacial cycle, the mid-Holocene and the present day. But the great majority, perhaps all of them, were caused by predation.

The impact of the arrival of people 45 kyr ago was overwhelmingly that of a new predator. The human population of the time was probably small, and people may not have specialised in large-mammal hunting. But the addition of a new and versatile predator to almost all the major environments of Greater Australia added a small but crucial mortality burden to populations of demographically susceptible large mammals. The result was the gradual decline and extinction of all the largest species. The stepping up of hunting pressure on mammals as a result of the social and technological changes of Aboriginal 'intensification' brought about another round of declines in the mid- to late-Holocene, and this may have been the cause of the mainland extinctions of the thylacine and the devil. (And if predation by people was not responsible for this, the arrival of another predator, the dingo, almost certainly was.) The decline of the thylacine under human predation was pushed to its conclusion under a deliberate program of hunting by British settlers in Tasmania.

The very high predation pressures imposed by cats and foxes were primarily responsible for a succession of waves of extinction that swept over the mainland during the last 200 years, following the invasions and subsequent population increases of those two predators. Cats and foxes continue to threaten native mammals, and might well cause more extinctions in the future. The devastating impact of new predators, and the pervasive effects of shifts in the balance of existing predator–prey interactions, are the themes that underlie the whole history of extinctions in Australia for the last 50 000 years.

But mammals in other parts of the world have been challenged by new predators without, it seems, suffering as badly as Australia's mammals. Why were the impacts of new predators so dire on this continent? Broadly, there are three ways in which this might be explained. The predators might have been unusually potent; or the mammal fauna might have been intrinsically susceptible to any new predator; and finally, there might have been some particular characteristics of the way in which events unfolded in Australia, some unique historical contingencies that can explain why things ultimately turned out so badly here. These three alternatives are considered in turn below.

EXTRAORDINARY PREDATORS?

Within the last 50 kyr modern *Homo sapiens* spread over the whole of the globe from our origins in Africa. The arrival of people in new lands was followed almost invariably by extinction of the largest vertebrates that they found there. Among many other species, these included mammoths, giant deer and giant bears in Europe and North America, as well as ground sloths, giant armadillos and horses in the Americas; giant lemurs and aardvarks in Madagascar; moa and the giant Haast's eagle in New Zealand; hippos and elephants on islands in the Mediterranean, and so on (Martin & Steadman 1999).

In general, the pattern of 'first-contact' extinction in Australia looks very similar to events elsewhere in the world. The difference was that in Australia the extinctions cut deeper into the fauna, leaving nothing above 45 kilograms alive, whereas much larger mammals managed to survive in Europe and the Americas. However, this cannot be explained by invoking some peculiar ferocity of the first Australians. There is no reason to think that they were unusually inclined to kill large vertebrates, and as far as we know they had no weapons that were specially crafted for that purpose – unlike the early people of North America, who left many lethal-looking stone points as evidence of their presence at the time when very large mammals went extinct there.

The other predators that caused such damage later in history were certainly formidable hunters. The cunning and skill of the fox is proverbial, and the cat is 'the quickest to his mark of any creature' (from *Jubilate Agno* by Christopher Smart, 1763). But foxes and cats occur through much of the world, and although they can be destructive there are many places where they coexist with mammals analogous to the 'critical weight range' species that they hunted to extinction in Australia.

A FRAGILE FAUNA?

If the predators themselves were not exceptional, was there something that made Australian mammals especially susceptible to them? This has often been argued, and from several different perspectives. In the nineteenth and the early twentieth centuries, marsupials were generally considered to be constitutionally inferior to placental mammals. Even the fact that living marsupials are concentrated in the south of the globe, in South America and Australia, was interpreted as a sign of weakness: they had been driven into southern refuges by superior placental mammals that came (naturally) from the northern hemisphere. Wood Jones, an early twentieth-century authority on Australian marsupials (and a man deeply concerned for their plight) described it this way:

> they were driven by the better endowed Monodelphians [placental mammals] into the dead ends of the earth. The Monodelphians, with superior intelligence and better equipment, pushed them in front, and drove them by successive waves down the great American continent, until now they have a rather slender footing in the southern portions of the New World. The same pressure probably drove them across Asia, and thence to the land bridge now represented by the islands to the South and East of the Malayan chain … they were driven into Australia, which, at the time of their advent, was connected with other landmasses of the globe. (Jones 1923–25)

On this view, it is no surprise that when placental predators reached Australia they destroyed so many of their marsupial prey. But the idea that marsupials are intrinsically inferior to placental mammals is no longer widely held, and the history of the origin and spread of marsupials has been completely rewritten since Wood Jones's day (see Chapter 1). In any case, placental superiority, even if it were an established fact, does not fit the pattern of extinctions of the last 200 years. Rodents are the most successful group of placental mammals worldwide, yet in Australia rodents suffered even more badly at the teeth of cats and foxes than did marsupials. There is also the evidence referred to in Chapter 12 that the arrival of the wolf – a superior placental predator if ever there was one – in the form of the dingo caused few or no extinctions.

An alternative explanation for the vulnerability of Australia's mammal fauna is that by virtue of its long isolation, Australia is more like an island than a continent (Flannery 1994). Island faunas are typically highly susceptible to the impacts of any alien species because their indigenous species live in tightly co-evolved communities that are prone to collapse if disturbed. They are especially vulnerable to new and unfamiliar predators, against which they have few or no defences. The view of Australia as a biological island with a naive fauna could help to explain why so many species from Australia's most ancient and endemic mammal lineages – monotremes, marsupials, conilurine rodents – seemed to simply vanish when new predators from other worlds arrived, while groups that had done most of their evolution elsewhere – bats and the more recently arrived rodents – survived unscathed.

However, there are problems with this explanation. First, as I pointed out in Chapter 3, naivety is impossible to verify for Pleistocene mammals and must simply be assumed. Second, although Australia did not have people before 45 kyr ago, or cats and foxes before 200 years ago, the continent had had its own mammalian predators for millions of years. It is therefore difficult to argue that the species that vanished in the wake of people, cats and foxes succumbed because they had never evolved the capacity to fear and avoid mammal predators, or to defend themselves if attacked. And finally, we can point to dingoes on the mainland and cats in Tasmania as cases where the arrival of novel predators did not cause collapse of prey populations.

Another possibility is that the vulnerability of the Australian mammal fauna was in some way a product of special features of the Australian environment itself. One idea that has run through the literature on Australian ecology is that Australian environments are uniquely fragile because of the continent's generally shallow and infertile soils, its variable climate with mostly low rainfall, and (as a consequence of these factors) its low energy availability. These environmental conditions might have put mammals in an especially precarious position.

Being endotherms, mammals have high energy requirements and need a regular supply of food. In environments with low and variable energy supply, mammals may naturally have had a somewhat tenuous grip on survival. This seems most self-evident for arid habitats, where reptiles are much more diverse than small mammals. As ectotherms, reptiles have much lower energy demands than mammals of similar size and so can maintain larger populations and tolerate long periods when little or no food is available (Morton 1990b, 1993). Further, environmental variability in space poses special problems for non-flying mammals. Birds are also endotherms and need a lot of energy, but in the arid zone they can move around to keep up with shifts in the locations of resources as rainfall varies from place to place. Bats can also do this, but non-flying mammals are tied down to the landscape and so are much more at the mercy of the harsh and unpredictable climate.

If Australian environments really are fundamentally unsuitable for non-flying mammals we might expect that any added pressure on them, whether from introduced competitors, environmental deterioration or increased predation, would push many species to extinction. Morton (1990a) developed an explicit model that used this idea to explain collapses of critical-weight-range mammals in arid Australia. He argued that small patches of relatively productive habitat were crucial for the persistence of many such mammals. The extinction of so many is explained by the damage done to those crucial patches by introduced herbivores.

Morton's hypothesis was persuasively argued for the arid zone, but it is not clear that it can explain why the same pattern of extinction was repeated in different environments, such as the open woodlands of south-eastern Australia. Johnson *et al.* (2005) tried to test the hypothesis by looking for demographic evidence of patch-dependence in two surviving critical-weight-range mammals from dry subtropical woodlands, but could find none. More generally, it is hard to believe that mammals were engaged in an uphill battle for survival in Australian environments when one considers the spectacular diversity of the late Pleistocene megafauna.

There is, however, one way in which evolution in Australia's environments might have put native mammals at a disadvantage when faced with exotic predators. One of the consistent traits of many Australian mammals is that they have low reproductive rates compared with mammals elsewhere (Tyndale-Biscoe 2005). Marsupial and placental mothers expend about the same total amount of energy to produce an offspring, but they schedule this energy expenditure differently. In marsupials, energy transfer from mother to young is spread over a longer period, and the peak level of energy transfer is correspondingly lower than in comparable placentals. Marsupial mothers therefore take longer to produce each of their offspring. This makes sense in the Australian environment, where energy availability is often low (Tyndale-Biscoe 2005), and it is especially obvious in herbivores, which have a naturally low-energy diet (Fisher *et al.* 2001).

It seems that the more time a mammal group has spent evolving in Australia, the more slowly do its members reproduce. Monotremes have been here the longest, and they have exceptionally slow reproduction (relative to body size); marsupials have slower reproduction than most placentals, but not as slow as monotremes; conilurine rodents, which have been evolving in Australia for the last four million years or so, reproduce faster than marsupials but more slowly than rodents elsewhere and, in particular, more slowly than the rodent species that invaded Australia more recently.

Slow reproduction is a trait that makes species demographically susceptible to overhunting, because it is linked to low population growth rates (see Chapters 6, 12). A slow rate of reproduction made the megafauna vulnerable to predation by humans, and it probably increased the vulnerability of critical-weight-range marsupials and conilurine rodents to foxes and cats. Slow reproduction may go a long way to explaining why extinction has cut so deeply into the Australian mammal fauna.

This is most clear in the comparison of megafauna extinctions in Australia and other places (Johnson 2002). For example, in both Australia and the northern hemisphere prehistoric extinctions were concentrated on the larger mammals. In North America and Eurasia, deer and bovids (antelope, wild cattle and their relatives) were very likely to go extinct if larger than about 380 kilograms, but in Australia kangaroos went extinct above a body mass of about 32 kilograms. This is a huge difference in the apparent severity of the extinction event, but it is accounted for by the fact that at any given body mass kangaroos reproduce more slowly than northern hemisphere mammals. A 32 kilogram kangaroo reproduces at about the same rate as a 380 kilogram deer, so if reproductive rate is used as the predictor of extinction risk, the difference between the two groups of mammals disappears (see also Figure 6.3).

ACCIDENTS OF HISTORY

Finally, it might be that there was some unique combination of events in Australia, some idiosyncratic but exceptionally harmful sequence of causes and consequences, that resulted in a large total effect on native mammals of humans and the other predators they introduced to this continent. I believe this is so, and that the sequence of events that led directly to Australia's high rate of recent mammals extinction began in prehistory.

This lethal sequence of events began with the simplification of Australia's community of large predators. Pleistocene Australia had a complex community of carnivores, including large mammals from the marsupial lion *Thylacoleo* down, as well as some large carnivorous reptiles. With the extinction of the largest species on human arrival, people stepped into the role of Australia's largest predator. But for many thousands of years they shared the continent with two other large carnivores, one of them a specialist predator of large mammals (the thylacine) and the other a predator and specialised scavenger (the devil). This changed in the mid-Holocene when – either because of the arrival of the dingo or, more probably, of increased human impact – the thylacine and devil went extinct on the mainland. The dingo took over the ecological roles of both the thylacine and devil, but the dingo itself was probably kept under tight ecological control by the human population. This subordination of the dingo would have increased as the human population, and the impact on mammal populations of human hunting, increased steadily over the last few thousand years of prehistory. By this time people had become by far the most significant predators on mainland Australia.

So, the complex Pleistocene community of large carnivores had been reduced to one consisting only of people and dingoes. Large top predators can have very important effects in reducing populations of smaller and potentially more destructive mesopredators (see Chapter 12), and, in the state that Australia had reached in late prehistory, it was very susceptible to invasion by such mesopredators. Those that arrived were cats and foxes, and vast areas were left completely open to them when Aborigines left most of their land as a consequence of the expansion of European pastoralism, and dingoes, having at first increased in abundance, were widely suppressed by the new settlers. The impact of these predators on their mammal prey was amplified by the fact that the European settlers had just introduced rabbits as well. Matters were made worse by the fires, vegetation clearing and livestock grazing that came with European land management and that left native prey species highly exposed to predation. This was a diabolical succession of events for medium-sized mammals of inland habitats, because at every step their ultimate exposure to predation by cats and foxes was increased. I believe that the peculiar features of this set of changes, acting on a fauna that was demographically sensitive to overpredation, explains why Australia has lost so many mammals, especially in the recent past. The same forces could cause more extinctions in the future.

How are we to set about reversing this ecological transformation of Australia? My suggestions at the end of Chapter 12 are that we rebuild predator communities in Australia to the extent possible, by making use of the dingo, and by trying to re-create landscapes that can reduce predation pressure on native species from cats and foxes. The dingo is probably our best hope for resetting the balance of predator–prey interactions in a way that gives small ground-dwelling mammals a chance at survival. But, there is one other step we could take that might help to achieve this.

We know that Tasmanian devils can coexist with species of native mammals that are highly susceptible to fox and cat predation, and that survive in Tasmania where devils are abundant. And we know that until recently, perhaps as little as 400 years ago, there were still devils on mainland Australia. Devils are tough, bold, truculent and well armed. They would probably not prey on adult foxes or cats, but a healthy population of devils might provide enough competition over carcasses, or kill cubs often enough, to prevent foxes becoming so common that they threatened critical-weight-range mammals. It is even possible that it has been the presence of devils that until recently prevented foxes from establishing in Tasmania (Menna Jones, personal communication; this is a thought that makes doubly alarming the recent decline in devil numbers due to disease). Perhaps a healthy population of devils might also help to suppress cats, and devils could be part of the explanation for the failure of feral cats to cause extinctions in Tasmania.

The reintroduction of Tasmanian devils into parts of southeastern Australia could help to build some resistance against introduced predators back into the mammal communities of the mainland (Wroe and Johnson 2004). It could be planned as part of a major program to return to the mainland other mammals that have gone extinct more recently and that survive in Tasmania: the red-bellied pademelon, Tasmanian bettong and eastern quoll. Reintroducing these four species would represent a significant recovery of mammal diversity to mainland Australia. And returning the largest living marsupial carnivore to the mainland would take us one small step back towards the lost richness and beauty of the mammal fauna of Pleistocene Australia.

REFERENCES

Abbott, I. (2001). The bilby *Macrotis lagotis* (Marsupialia: Peramelidae) in south-western Australia: original range limits, subsequent decline, and presumed regional extinction. *Records of the Western Australian Museum*, **20**, 271–305.

—— (2002). Origin and spread of the cat, *Felis catus*, on mainland Australia, with a discussion of the magnitude of its early impact on native fauna. *Wildlife Research*, **29**, 51–74.

—— (2003). Aborigines, settlers and native animals: a zoological history of the south-west. *Early Days*, **12**, 231–49.

Akcakaya, H.R. (2002). RAMAS Metapop: viability analysis for stage-structured metapopulations (version 4.0). Applied Biomathematics, Setauket, New York.

Akerman, K. (1973). Two Aboriginal charms incorporating giant marsupial teeth. *Western Australian Naturalist*, **12**, 139–41.

—— (1998). A rock painting, possibly of the now extinct marsupial *Thylacoleo* (marsupial lion), from the north Kimberley, Western Australia. *The Beagle: Records of the Museums and Art Galleries of the Northern Territory*, **14**, 117–21.

Allan, G.E. & Southgate, R.I. (2002). Fire regimes in the spinifex landscapes of Australia, in *Flammable Australia: the fire regimes and biodiversity of a continent* (eds R.A. Bradstock, J.E. Williams & A.M. Gill), pp. 199-237. Cambridge University Press, Cambridge.

Alroy, J. (1999). Putting North America's end-Pleistocene megafaunal extinction in context: large-scale analyses of spatial patterns, extinction rates, and size distributions, in *Extinctions in Near Time: causes, contexts and consequences* (ed R.D.E. MacPhee), pp. 105–44. Kluwer Academic/Plenum Publishers, New York.

—— (2001). A multispecies overkill simulation of the end-Pleistocene megafaunal mass extinction. *Science*, **292**, 1893–96.

Anderson, E. (1993). *Plants of central Queensland: their identification and uses* Queensland, Department of Primary Industries, Brisbane.

Aplin, K., Pasveer, J.M. & Boles, W.E. (1999). Late Quaternary vertebrates from the Bird's Head Peninsula, Irian Jaya, Indonesia, including descriptions of two previously unknown marsupial species. *Records of the Australian Museum*, **supplement 57**, 351–87.

Archer, M. (1974). New information about the Quaternary distribution of the thylacine (Marsupialia, Thylacinidae) in Australia. *Journal of the Royal Society of Western Australia*, **57**, 43–9.

—— (1984). Effects of humans on the Australian vertebrate fauna. In *Vertebrate Zoogeography and Evolution in Australia* (eds M. Archer & G. Clayton), pp. 151–61. Hesperian Press, Perth.

Archer, M. & Baynes, A. (1972). Prehistoric mammal faunas from two small caves in the extreme south-west of Western Australia. *Journal of the Royal Society of Western Australia*, **55**, 80–9.

Archer, M., Crawford, I.M., & Merrilees, D. (1980). Incisions, breakages and charring, some probably man-made, in fossil bones from Mammoth Cave, Western Australia. *Alcheringa*, **4**, 115–31.

Archer, M., Godthelp, H. & Hand, S.J. (1993). Early Eocene marsupials from Australia. *Kaupia*, **3**, 193–200.

Archer, M. *et al.* (1999). The evolutionary history and diversity of Australian mammals. *Australian Mammalogy*, **21**, 1-45.

Attenbrow, V. (2002). *Sydney's Aboriginal Past*. University of New South Wales Press, Sydney.

Ayliffe, L.K. *et al.* (1998). 500 ka precipitation record from southeastern Australia: evidence for interglacial relative aridity. *Geology*, **26**, 147–50.

Baird, R.F. (1985). *Centropus* the giant coucal, in *Kadimakara: extinct vertebrates of Australia* (eds P.V. Rich & G.F. van Tets), pp. 205–7. Pioneer Design Studio, Melbourne.

Balme, J. (2000). Excavations revealing 40,000 years of occupation at Mimbi Caves, south central Kimberley, Western Australia. *Australian Archaeology*, **51**, 1–5.

Balme, J. & Hope, J. (1990). Radiocarbon dates from midden sites in the lower Darling River area of western New South Wales. *Archaeology in Oceania*, **25**, 85–101.

Balme, J., Merrilees, D. & Porter, J.K. (1978). Late Quaternary mammal remains, spanning about 30 000 years, from excavations in Devil's Lair, Western Australia. *Journal of the Royal Society of Western Australia*, **61**, 33–65.

Banks, P.B., Dickman, C.R., & Newsome, A.E. (2000). Predation by red foxes limits recruitment in populations of eastern grey kangaroos. *Austral Ecology*, **25**, 283–91.

Barker, R.D. & Caughley, G. (1990). Distribution and abundance of kangaroos (Marsupialia: Macropodidae) at the time of European contact: Tasmania. *Australian Mammalogy*, **13**, 157–66.

Barker, R.D. & Caughley, G. (1992). Distribution and abundance of kangaroos (Marsupialia: Macropodidae) at the time of European contact: Victoria. *Australian Mammalogy*, **15**, 81–8.

Barker, W.R., Barker, R.M., & Haegi, L. (1999). Introduction to Hakea, in *Flora of Australia*, vol. 17 B, *Proteaceae 3: Hakea to Dryandra*, pp. 1–30. Australian Biological Resources Study, Canberra.

Barlow, C. (2000). *The ghosts of evolution*. Basic Books, New York.

Barrows, T.T. & Juggins, S. (2005). Sea-surface temperatures around the Australian margin and Indian Ocean during the Last Glacial Maximum. *Quaternary Science Reviews*, **24**, 1017–47.

Barrows, T.T., Stone, J.O., Fifield, L.K. & Cresswell, R.G. (2001). Late Pleistocene glaciation of the Kosciuszko massif, Snowy Mountains, Australia. *Quaternary Research*, **55**, 179-189.

Belovsky, G.E., Schmitz, O.J., Slade, J.B. & Dawson, T.J. (1991). Effects of spines and thorns on Australian arid zone herbivores of different body masses. *Oecologia*, **88**, 521–8.

Benson, J.S. & Redpath, P.A. (1997). The nature of pre-European native vegetation in south-eastern Australia: a critique of Ryan D.G., Ryan, J.R. and Starr, B.J. (1995) *The Australian Landscape: observations of explorers and early settlers*. *Cunninghamia*, **52**, 285–328.

Bird, M.I. *et al.* (2002). Radiocarbon analysis of the early archaeological site of Nauwalabila I, Arnhem Land, Australia: implications for sample suitability and stratigraphic integrity. *Quaternary Science Reviews*, **21**, 1061–75.

Bishop, N. (1997). Functional anatomy of the macropodid pes. *Proceedings of the Linnean Society of New South Wales*, **117**, 17–50.

Blumstein, D.T. & Daniel, J.C. (2003). Isolation from mammalian predators differentially affects two congeners. *Behavioral Ecology*, **13**, 657–63.

Blunier, T. & Brook, E. (2001). Timing of millennial-scale climate change in Antarctica and Greenland during the last glacial period. *Science*, **291**, 109–12.

Boeda, E., Geneste, J.M. & Griggo, C. (1999). A Levallois point embedded in the vertebra of a wild ass (*Equus africanus*): hafting, projectiles and Mousterian hunting weapons. *Antiquity*, **73**, 394–402.

Bolton, B.L. & Latz, P.K. (1978). The western hare-wallaby *Lagorchestes hirsutus* (Gould) (Macropodidae) in the Tanami Desert. *Australian Wildlife Research*, **5**, 285–93.

Bond, W.J., Lee, W.G. & Craine, J.M. (2004). Plant structural defences against browsing birds: a legacy of New Zealand's extinct moas. *Oikos*, **104**, 500–8.

Bowdler, S. (1977). The coastal colonisation of Australia. In *Sunda and Sahul: prehistoric studies in southeast Asia, Melanesia and Australia* (eds J. Allen, J. Golson & R. Jones). Academic Press, London.

Bowler, J.M. (1982). Aridity in the late tertiary and Quaternary of Australia, in *Evolution of the flora and fauna of arid Australia* (eds W.R. Barker & P.J.M. Greenslade), pp. 354–5. Peacock Publications, Adelaide.

Bowler, J.M. (1998). Willandra Lakes revisited: environmental framework for human occupation. *Archaeology in Oceania*, **33**, 12 055.

Bowler, J.M. & Magee, J.W. (2000). Redating Australia's oldest human remains: a sceptic's view. *Journal of Human Evolution*, **38**, 719–26.

Bowler, J.M., Wyrwoll, K.-H. & Lu, Y. (2001). Variations of the northwest Australian summer monsoon over the last 300,000 years: the paleohydrological record of the Gregory (Mulan) Lakes System. *Quaternary International*, **83–85**, 63–80.

Bowler, J.M. *et al.* (2003). New ages for human occupation and climatic change at Lake Mungo, Australia. *Nature*, **421**, 837–40.

Bowman, D.M.J.S. (1998). Tansley Review no. 101: The impact of Aboriginal landscape burning on the Australian biota. *New Phytologist*, **140**, 385–410.

—— (2000). *Australian rainforests: islands of green in a land of fire*. Cambridge University Press, Cambridge.

—— (2003). Australian landscape burning: a continental and evolutionary perspective, in *Fire in Ecosystems of South-west Western Australia* (eds I. Abbott & N. Burrows), pp. 107–18. Backhuys Publishers, Leiden, the Netherlands.

Bowman, D.M.J.S., Garde, M. & Saulwick, A. (2001). Kunj-ken Makka Man-wurrk Fire is for kangaroos: interpreting Aboriginal accounts of landscape burning in central Arnhem Land, in *Histories of Old Ages: essays in honour of Rhys Jones* (eds A. Anderson, I. Lilley & S. O'Connor), pp. 61–78. Pandanus Books, Canberra.

Bowman, D.M.J.S. & Prior, L.D. (2004). Impact of Aboriginal landscape burning on woody vegetation in *Eucalyptus tetrodonta* savanna in Arnhem Land, northern Australia. *Journal of Biogeography*, **31**, 807–17.

Bowman, D.M.J.S., Walsh, A. & Prior, L.D. (2004). Landscape analysis of Aboriginal fire management in Central Arnhem Land, north Australia. *Journal of Biogeography*, **31**, 1–17.

Braithwaite, R.W. & Griffiths, A.D. (1994). Demographic variation and range contraction in the northern quoll, *Dasyurus hallucatus* (Marsupialia: Dasyuridae). *Wildlife Research*, **21**, 203–17.

Breed, W. & Ford, F. (2006). *Native Rats And Mice.* CSIRO Publishing, Melbourne.

Brook, B.W. & Bowman, D.M.J.S. (2002). Explaining the Pleistocene megafaunal extinctions: models, chronologies, and assumptions. *Proceedings of the National Academy of Science of the USA*, **99**, 14 624–7.

—— (2004). The uncertain blitzkrieg of Pleistocene megafauna. *Journal of Biogeography*, **31**, 517–23.

Brook B.W. & Johnson, C.N. (2006). Selective hunting of juveniles as a cause of the imperceptible overkill of the Australasian Pleistocene 'megafauna'. *Alcheringa* (in press).

Browne, W.R. (1945). President's address. *Proceedings of the Linnean Society of New South Wales*, **70**, I–xxv.

Bulmer, S. (2001). Lapita dogs and singing dogs and the history of the dog in New Guinea, in *The Archaeology of Lapita Dispersal in Oceania* (eds G.R. Clark, A.J. Anderson & T. Vunidilo). Pandanus Books, Canberra.

Bulte, E.H., Horan, R.D. & Shogren, J.F. (2003). Is the Tasmanian tiger extinct?: a biological-economic re-evaluation. *Ecological Economics*, **45**, 271–9.

Burbidge, A.A., Johnson, K.A., Fuller, P.J. & Southgate, R.J. (1988). Aboriginal knowledge of the mammals of the central deserts of Australia. *Australian Wildlife Research*, **15**, 939.

Burbidge, A.A. & Mckenzie, N.L. (1989). Patterns in the modern decline of Western Australia's vertebrate fauna: causes and conservation implications. *Biological Conservation*, **50**, 143–98.

Burbidge, A.A. & Manly, B.J.F. (2002). Mammal extinctions on Australian islands: causes and conservation implications. *Journal of Biogeography*, **29**, 465–73.

Burbidge, N.T. (1960). The phytogeography of the Australian region. *Australian Journal of Botany*, **8**, 75–212.

Burness, G.P., Diamond, J. & Flannery, T. (2001). Dinosaurs, dragons, and dwarfs: the evolution of maximal body size. *Proceedings of the National Academy of Science of the USA*, **98**, 14 518–23.

Butlin, N.G. (1962). Distribution of the sheep population: preliminary statistical picture, 1860–1957. In *The Simple Fleece: studies in the Australian wool industry* (ed J.A. Barnard), pp. 281–307. Melbourne University Press, Melbourne.

Calaby, J.H. & White, C. (1967). The Tasmanian devil (*Sarcophilus harrisii*) in northern Australia in recent times. *Australian Journal of Science*, **29**, 473–5.

Cane, S. (2001). The great flood: eustatic change and cultural change in Australian during the Late Pleistocene and Holocene. In *Histories of Old Ages: essays in honour of Rhys Jones* (eds A. Anderson, I. Lilley & S. O'Connor), pp. 141–66. Pandanus Books, Canberra.

Cardillo, M. (2003). Biological determinants of extinction risk: why are smaller species less vulnerable? *Animal Conservation*, **6**, 63–9.

Cardillo, M., Bininda-Edmonds, O.R.P., Boakes, E. & Purvis, A. (2004). A species-level phylogenetic supertree of marsupials. *Journal of Zoology* (London), **264**, 11–31.

Cardillo, M. & Bromham, L. (2001). Body size and risk of extinction in Australian mammals. *Conservation Biology*, **15**, 1435–49.

Carr, S.G. & Robinson, A.C. (1997). The present status and distribution of the desert rat-kangaroo *Caloprymnus campestris* (Marsupialia: Potoroidae). *South Australian Naturalist*, **72**, 4–27

Caughley, G. (1977). *Analysis of Vertebrate Populations*. John Wiley & Sons, Chichester UK.

Caughley, G. & Gunn, A. (1996). *Conservation Biology in Theory and Practice*, Blackwell Science, Cambridge, Mass.

Cayzer, L.W., Crisp, M.D., & Telford, I.R.H. (1999). *Bursaria* (Pittosporaceae): a morphometric analysis and revision. *Australian Systematic Botany*, **12**, 117–43.

—— (2000). Revision of Pittosporum (Pittosporaceae) in Australia. *Australian Systematic Botany*, **13**, 845–902.

Cerling, T.E., Harris, J.M., McFadden, B.J., Leakey, M.G., Quade, J., Eisenmann, V. & Ehleringer, J.R. (1997) Global vegetation change through the Miocene/Pliocene boundary. *Nature*, **389**, 153–8.

Chaloupka, G. (1993). *Journey in Time*. Reed Books, Sydney.

Chappell, J. (2002). Sea level changes forced ice breakouts in the Last Glacial cycle: new results from coral terraces. *Quaternary Science Reviews*, **21**, 1229–40.

Chivas, A.R. *et al.* (2001). Sea-level and environmental changes since the last interglacial in the Gulf of Carpentaria, Australia: an overview. *Quaternary International*, **83–85**, 19–46.

Choquenot, D. & Bowman, D.M.J.S. (1998). Marsupial megafauna, Aborigines and the overkill hypothesis: application of predator-prey models to the question of Pleistocene extinction in Australia. *Global Ecology and Biogeography Letters*, **7**, 167–80.

Christensen, P.E.S. (1980). The biology of *Bettongia penicillata* Gray, 1837 and *Macropus eugenii* (Desmarest, 1817) in relation to fire, Rep. no. 91. Forests Department of Western Australia.

Cifelli, R.L. & Davis, B.M. (2003). Marsupial origins. *Science*, **302**, 1899–1900.

Cifelli, R.L. & Muizon, C.D. (1997). Dentition and jaw of *Kokopellia juddi*, a primitive marsupial or near-marsupial from the medial Cretaceous of Utah. *Journal of Mammalian Evolution*, **4**, 241–58.

Claridge, A.W. & May, T.W. (1994). Mycophagy among Australian mammals. *Australian Journal of Ecology*, **19**, 251–5.

Claridge, A.W. *et al.* (1992). Establishment of ectomycorrhizae on the roots of two species of *Eucalyptus* from fungal spores contained in the faeces of the long-nosed potoroo (*Potorous tridactylus*). *Australian Journal of Ecology*, **17**, 207–17.

Clement, A.C., Seager, R. & Cane, M.A. (1999). Orbital controls on the El Niño /Southern Oscillation and the tropical climate. *Paleoceanography*, **14**, 441–56.

Colhoun, E.A. (2000). Vegetation and climate change during the Last Interglacial-Glacial cycle in western Tasmanian, Australia. *Palaeogeography, Palaeoclimatology, Palaeoecology*, **155**, 195–209.

Colhoun, E.A., Pola, J.S., Barton, C.E. & Heijnis, H. (1999). Late Pleistocene vegetation and climate history of Lake Selina, western Tasmania. *Quaternary International*, **57/58**, 5–23.

Colhoun, E.A. & van de Geer, G. (1988). Darwin Crater, the King and Linda valleys, in *Cainozoic Vegetation of Tasmania: handbook prepared for the 7th International Palynological Congress* (ed E.A. Colhoun), pp. 30–71. Department of Geography, University of Newcastle, Newcastle.

Coltrain, J.B., Field, J., Cosgrove, R. & O'Connell, J. (2004). Stable isotope and protein analyses of Cuddie Springs *Genyornis*. *Archaeology in Oceania*, **39**, 50–1.

Cooke, B. & Kear, B. (1999). Evolution and diversity of kangaroos (Macropodidae, Marsupialia). *Australian Mammalogy*, **21**, 27–9.

Cooke, B.D. & Fenner, F. (2002). Rabbit haemorrhagic disease and the biological control of wild rabbits, *Oryctolagus cuniculus*, in Australia and New Zealand. *Wildlife Research*, **29**, 689–706.

Cooper, S.M. & Owen-Smith, N. (1986). Effects of plant spinescence on large mammalian herbivores. *Oecologia*, **68**, 446–55.

Corbett, L. (1995). *The Dingo in Australia and Asia*. University of New South Wales Press, Sydney.

Cosgrove, R. (1995). Late Pleistocene behavioural variation and time trends: the case from Tasmania. *Archaeology in Oceania*, **30**, 83–104.

—— (1996). Origin and development of Australian Aboriginal tropical rainforest culture: a reconsideration. *Antiquity*, **70**, 900–12.

Cosgrove, R. & Allen, J. (2001). Prey choice and hunting strategies in the Late Pleistocene: evidence from southwest Tasmania, in *Histories of Old Ages: essays in honour of Rhys Jones* (eds A. Anderson, I. Lilley & S. O'Connor), pp. 397–430. Pandanus Books, Canberra.

Courchamp, F., Langlais, M. & Sugihara, G. (1999). Cats protecting birds: modelling the mesopredator release effect. *Journal of Animal Ecology*, **68**, 282–92.

—— (2000). Rabbits killing birds: modelling the hyperpredation process. *Journal of Animal Ecology*, **69**, 154-164.

Courtenay, J.M. & Friend, J.A. (2004). Gilbert's potoroo recovery plan, July 2003 – June 2008, Rep. no. 32. Western Australia, Department of Conservation and Land Management, Perth.

Crooks, K.R. & Soule, M.E. (1999). Mesopredator release and avifaunal extinctions in a fragmented system. *Nature*, **400**, 563–6.

Cunningham, G.M., Mulham, E.E., Milthorpe, P.L. & Leigh, J.H. (1981) *Plants of Western New South Wales*. New South Wales Government Printing Office, Sydney.

David, B. (2002). *Landscapes, Rock-art and the Dreaming*. Leicester University Press, London.

David, B. & Lourandos, H. (1997a). 37,000 years and more in tropical Australia: investigating long-term archaeological trends in Cape York Peninsula. *Proceedings of the Prehistoric Society*, **63**, 1–23.

David, B. & Lourandos, H. (1998). Rock art and socio-demography in northeastern Australian prehistory. *World Archaeology*, **30**, 193–219.

David, B. *et al.* (1997b). New optical and radiocarbon dates from Ngarrabullgan Cave, a Pleistocene archaeological site in Australia: implications for the comparability of time clocks and for the human colonization of Australia. *Antiquity*, **71**, 183–8.

Davidson, D.S. (1936). The spearthrower in Australia. *Proceedings of the American Philosophical Society*, **16**, 445–83.

Dawson, L. (1985). Marsupial fossils from Wellington Caves, New South Wales: the historic and scientific significance of the collections in the Australian Museum, Sydney. *Records of the Australian Museum*, **37**, 55–69.

D'Costa, D.M. (1997). The reconstruction of Quaternary vegetation and climate on King Island, Bass Strait, Australia. PhD thesis, Monash University, Melbourne.

De Deckker, P. (2001). Late Quaternary cyclic aridity in tropical Australia. *Palaeogeography, Palaeoclimatology, Palaeoecology*, **170**, 1–9.

DeVogel, S.B., Magee, J.W., Manley, W.F. & Miller, G.H. (2004) A GIS-based reconstruction of Late Quaternary paleohydrology: Lake Eyre, arid central Australia. *Palaeogeography, Palaeoclimatology, Palaeoecology*, **204**, 1–13.

Dickman, C.R. (1996a). Impact of exotic generalist predators on the native fauna of Australia. *Wildlife Biology*, **2**, 185–95.

—— (1996b). *Overview of the Impacts of Feral cats on Australian Native Fauna* Australian Nature Conservation Agency, Canberra.

—— (1996c). Incorporating science into recovery planning for threatened species, in *Back from the Brink: refining the threatened species recovery process* (eds S. Stephens & S. Maxwell), pp. 63–73. Surrey Beatty & Sons, Sydney.

Dickman, C.R., Pressey, R.L., Lim, L. & Parnaby, H.E. (1993). Mammals of particular conservation concern in the western division of New South Wales. *Biological Conservation*, **65**, 219–48.

Dodson, J. & Mooney, S.D. (2002). An assessment of historic human impact on south-eastern Australian environmental systems, using late Holocene rates of environmental change. *Australian Journal of Botany*, **50**, 455–64.

Duncan, J. (2001). Megafauna at Keilor and the timing of their extinction. *Australian Archaeology*, **53**, 16–22.

Edwards, D.A. & O'Connell, J.F. (1995). Broad spectrum diets in arid Australia. *Antiquity*, **69**, 769–83.

Edwards, G.P., Dobbie, W. & Berman, D.M. (2002). Population trends in European rabbits and other wildlife of central Australia in the wake of rabbit haemorrhagic disease. *Wildlife Research*, **29**, 689–706.

Edwards, W., Gadek, P., Weber, E. & Worboys, S. (2001). Idiosyncratic phenomenon of regeneration from cotyledons in the idiot fruit tree, *Idiospermum australiense*. *Austral Ecology*, **26**, 254–8.

Ehleringer, J.R. & Monson, R.K. (1993). Evolutionary and ecological aspects of photosynthetic pathway variation. *Annual Review of Ecology and Systematics*, **24**, 411–39.

Ellis, C.J. (1997). Factors influencing the use of stone projectile tips: an ethnographic perspective, in *Projectile technology* (ed. H. Knecht), pp. 37–78. Plenum Press, New York.

EPICA (2004). Eight glacial cycles from an Antarctic ice core. *Nature*, **429**, 623–8.

Fethney, J., Roman, D. & Wright, R.V.S. (1987). Uranium series dating of Diprotodon teeth from archaeological sites on the Liverpool Plains, in *Proceedings of the Fifth Australian Conference on Nuclear Techniques of Analysis*, pp. 24–6.

Field, J. & Dodson, J. (1999). Late Pleistocene megafauna and archaeology from Cuddie Springs, south-eastern Australia. *Proceedings of the Prehistoric Society*, **65**, 275–301.

Field, J.H., Dodson, J.R. & Prosser, I.P. (2002). A late Pleistocene vegetation history from the Australian semi-arid zone. *Quaternary Science Reviews*, **21**, 1023–37.

Field, J. & Fullagar, R. (2001). Archaeology and Australian megafauna. *Science*, **294**, 7a.

Fifield, L.K. *et al.* (2001). Radiocarbon dating of the human occupation of Australia prior to 40 ka BP: successes and pitfalls. *Radiocarbon*, **43**, 1139–45.

Finke, D.L. & Denno, R.F. (2004). Predator diversity dampens trophic cascades. *Nature*, **429**, 407–10.

Finlayson, C. (2005). Biogeography and evolution of the genus Homo. *Trends in Ecology and Evolution*, **20**, 457–63.

Finlayson, H.H. (1927). Observations on the South Australian members of the subgenus 'Wallabia'. *Transactions of the Royal Society of South Australia*, **51**, 363–77.

Finlayson, H.H. (1932). *Caloprymnus campestris*, its recurrence and characters. *Transactions of the Royal Society of South Australia*, **56**, 146–7.

Finlayson, H.H. (1935a). *The Red Centre: man and beast in the heart of Australia*. Angus & Robertson Ltd, Sydney.

—— (1935b). On mammals from the Lake Eyre Basin Part II: the Peramelidae. *Transactions of the Royal Society of South Australia*, **59**, 227–36.

—— (1958). On central Australian mammals, Part III: the Potoroinae. *Records of the South Australian Museum*, **13**, 235–303.

—— (1961). On central Australian mammals, Part IV: the distribution and status of central Australian species. *Records of the South Australian Museum*, **14**, 141–91.

Fisher, D.O., Blomberg, S.O. & Owens, I.P.F. (2003). Extrinsic vs intrinsic factors in the decline and extinction of Australian marsupials. *Proceedings of the Royal Society of London B*, **270**, 1801–8.

Fisher, D.O., Owens, I.P.F. & Johnson, C.N. (2001). The ecological basis of life history variation in marsupials. *Ecology*, **82**, 3531–40.

Flannery, T.F. (1982). Hindlimb structure and evolution in the kangaroos (Marsupialia: Macropodidae), in *The Fossil Vertebrate Record of Australasia* (eds P.V. Rich & E.M. Thompson), pp. 507–24. Monash University, Melbourne.

—— (1990). Pleistocene faunal loss: implications of the aftershock for Australia's past and future. *Archaeology in Oceania*, **25**, 45–67.

—— (1994). *The Future Eaters*. Reed Books, Melbourne.

—— (1995). *Mammals of the South-West Pacific and Moluccan Islands*. Reed Books, Sydney.

—— (1999). The Pleistocene mammal fauna of Kelangurr Cave, central montane Irian Jaya, Indonesia. *Records of the Western Australian Museum*, **supplement 57**, 341–50.

—— (2004). *Country*. Text Publishing, Melbourne.

Flannery, T.F., Archer, M., Rich, T.H. & Jones, R. (1995). A new family of monotremes from the Cretaceous of Australia. *Nature*, **377**, 418–20.

Flannery, T.F., Martin, R. & Szalay, A. (1996) *Tree Kangaroos, a curious natural history*. Reed Books, Melbourne.

Flannery, T.F., Mountain, M.J. & Aplin, K. (1983). Quaternary kangaroos (Macropodidae: Marsupialia) from Nombe Rock Shelter, Papua New Guinea, with comments on the nature of megafaunal extinction in the New Guinea Highlands. *Proceedings of the Linnean Society of New South Wales*, **107**, 75–97.

Fleay, D. (1932). The rare dasyures (Native Cats). *Victorian Naturalist*, **49**, 63–8.

Fleming, P., Corbett, L., Harden, R. & Thomson, P. (2001). *Managing the impacts of dingoes and other wild dogs* Bureau of Rural Science, Canberra.

Flood, J. (1973). Pleistocene human occupation and extinct fauna in Clogg's Cave, Buchan, southeastern Australia. *Nature*, **246**, 303.

—— (1974). Pleistocene man at Cloggs Cave: his tool kit and environment. *Mankind*, **9**, 175–88.

—— (1997). *Rock Art of the Dreamtime*. Angus & Robertson, Sydney.

—— (1999). *Archaeology of the Dreaming*. Angus & Roberston, Sydney.

Flood, J., David, B., Magee, J. & English, B. (1987). Birrigai: a Pleistocene site in the south-eastern highlands. *Archaeology in Oceania*, **22**, 9–26.

Forbes, M., Bestland, E. & Wells, R. (2004). Preliminary ^{14}C dates on bulk soil organic matter from the Black Creek megafauna fossil site, Rocky River, Kangaroo Island, South Australia. *Radiocarbon*, **46**, 437–43.

Francois, R. (2004). Cool stratification. *Nature*, **428**, 31–2.

Freeland, W.J. (1994). Parasites, pathogens and the impacts of introduced organisms on the balance of nature in Australia, in *Conservation biology in Australia and Oceania* (eds C. Moritz & J. Kikkawa), pp. 171–80. Surrey Beatty & Sons, Sydney.

Friend, J.A. (1990). The numbat *Myrmecobius fasciatus* (Myrmecobiidae): history of decline and potential for recovery, in *Australian Ecosystems: 200 years of utilization, degradation and reconstruction* (eds D.A. Saunders, A.J.M. Hopkins & R.A. How), pp. 369–77. Surrey Beatty & Sons, Sydney.

Fullagar, R. & Field, J. (1997). Pleistocene seed-grinding implements from the Australian arid zone. *Antiquity*, **71**, 300–7.

Gaffney, E.S. (1991). The fossil turtles of Australia, in *Vertebrate Palaeontology of Australasia* (eds P. Vickers-Rich, J.M. Monaghan, R.F. Baird & T.H. Rich), pp. 703–20. Pioneer Design Studio/Monash University Publications Committee, Melbourne.

Gagan, M.K. & Chappell, J. (2000). Massive corals: grand archives of ENSO, in *El Niño: history and crisis* (eds R.H. Groves & J. Chappell), pp. 35–50. White Horse Press, Cambridge.

Garkaklis, M.J., Bradley, J.S., & Wooller, R.D. (1998). The effect of woylie (*Bettongia penicillata*) foraging on soil water repellency and water infiltration in heavy textured soil in southwestern Australia. *Australian Journal of Ecology*, **23**, 492–6.

—— (2000). Digging by vertebrates as an activity promoting the development of water repellent patches in sub-surface soils. *Journal of Arid Environments*, **45**, 35–42.

—— (2004). Digging and soil turnover by a mycophagous marsupial. *Journal of Arid Environments*, **56**, 569–78.

Gehring, C.A., Wolf, J.E. & Theimer, T.C. (2002). Terrestrial vertebrates promote arbuscular mycorrhizal fungal diversity and inoculum potential in a rainforest soil. *Ecology Letters*, **5**, 540–8.

Geiser, F. (2004). The role of torpor in the life of Australian arid zone mammals. *Australian Mammalogy*, **26**, 125–34.

Gentilli, J. (1961). Quaternary climates of the Australian region. *Annals of the New York Academy of Sciences*, **95**, 465–501.

Gibson, D.F. (1986). A biological survey of the Tanami Desert in the Northern Territory., Rep. no. 30, Conservation Commission of the Northern Territory.

Gibson, D.F. *et al.* (1994). Predation by feral cats, *Felis catus*, on the rufous hare-wallaby, *Lagorchestes hirsutus*, in the Tanami Desert. *Australian Mammalogy*, **17**, 103–7.

Giles, E. (1889). *Australia Twice Traversed*. Sampson Low, Marston, Searle & Rivington, London.

Gill, E.D. (1953). Distribution of the Tasmanian devil, the Tasmanian wolf, and the dingo in S.E. Australia in Quaternary time. *Victorian Naturalist*, **70**, 86–90.

Gill, E.D. (1955). Problem of extinction with special reference to Australian marsupials. *Evolution*, **9**, 87–92.

Gillespie, R. (2002). Dating the first Australians. *Radiocarbon*, **44**, 455–72.

Gillespie, R. & Brook, B.W. (2006). Is there a Pleitocene archaeological site at Cuddie Springs? *Archaeology in Oceania*, **41**, 1–11.

Gillespie, R. & David, B. (2001). The importance, or impotence, of Cuddie Springs. *Australasian Science*, **22**, 42–3.

Gillespie, R. *et al.* (1978). Lancefield swamp and the extinction of the Australian megafauna. *Science*, **200**, 1044–8.

Gillespie, R. & Roberts, R.G. (2000). On the reliability of age estimates for human remains at Lake Mungo. *Journal of Human Evolution*, **38**, 727–32.

Glen, A. & Short, J. (2000). Control of dingoes in New South Wales in the period 1883–1930 and its likely impact on their distribution and abundance. *Australian Zoologist*, **31**, 432–42.

Glen, A.S. & Dickman, C.R. (2005). Complex interactions among mammalian carnivores in Australia, and their implications for wildlife management. *Biological Reviews*, **80**, 1–15.

Godthelp, H. *et al.* (1992). Earliest known Australian Tertiary mammal fauna. *Nature*, **356**, 514–16.

Goede, A. & Bada, J.L. (1985). Electron spin resonance dating of Quaternary bone material from Tasmanian caves: a comparison with ages determined by aspartic acid recemization and C14. *Australian Journal of Earth Sciences*, **32**, 155–62.

Goede, A. & Murray, P. (1977). Pleistocene man in south central Tasmania: evidence from a cave site in the Florentine Valley. *Mankind*, **11**, 2–10.

Goede, A., Murray, P. & Harmon, R. (1978). Pleistocene man and megafauna in Tasmania: dated evidence from cave sites. *The Artefact*, **3**, 139–49.

Goin, F.J. *et al.* (1999). New discoveries of 'opossum-like' marsupials from Antarctica (Seymour Island, medial Eocene). *Journal of Mammalian Evolution*, **6**, 335–65.

Golson, J. (2001). New Guinea, Australia and the Sahul Connection, in *Histories of Old Ages: essays in honour of Rhys Jones* (eds A. Anderson, I. Lilley & S. O'Connor), pp. 185–210. Pandanus Books, Canberra.

Gorecki, P., Horton, D.R., Stern, N. & Wright, R.V.S. (1984). Coexistence of humans and megafauna in Australia: improved stratigraphic evidence. *Archaeology in Oceania*, **19**, 117–19.

Gott, B. (1983). Murnong – *Microseris scapigera*: a study of a staple food of Victorian Aborigines. *Australian Aboriginal Studies*, **1983/82**, 2–17.

Gould, R.A., O'Connor, S.A. & Veth, P. (2002). Bones of contention: reply to Walshe. *Archaeology in Oceania*, **37**, 96–101.

Grayson, D.K. (2001). The archaeological record of human impacts on animal populations. *Journal of World Prehistory*, 15, 1–68.

Grayson, D.K. & Meltzer, D.J. (2003). A requiem for North American overkill. *Journal of Archaeological Science*, **30**, 585–93.

Green, R.H. (1973). *The Mammals of Tasmania*. Mary Fisher Bookshop, Launceston.

Gröcke, D.R. (1997). Distribution of C3 and C4 plants in the Late Pleistocene of South Australia recorded by isotope biogeochemistry of collagen in megafauna. *Australian Journal of Botany*, **45**, 607–17.

Groube, L., Chappell, J., Muke, J. & Price, D. (1986). A 40,000 year-old human occupation site at Huon Peninsula, Papua New Guinea. *Nature*, **324**, 453–5.

Grove, R. (2005). Revolutionary weather: the climatic and economic crisis of 1788–1795 and the discovery of El Niño, in *A change in the weather* (eds T. Sherratt, T. Griffiths & L. Robin), pp. 128–40. National Museum of Australia Press, Canberra.

Grubb, P.J. (1992). A positive distrust of simplicity: lessons from plant defences and from competition among plants and among animals. *Journal of Ecology*, **80**, 585–610.

Grün, R., Moriarty, K. & Wells, R.T. (2001). Electron spin resonance dating of the fossil deposits in the Naracoorte Caves, South Australia. *Journal of Quaternary Science*, **16**, 49–59.

Grün, R. *et al.* (2000). Age of the Lake Mungo 3 skeleton, reply to Bowler & Magee and to Gillespie & Roberts. *Journal of Human Evolution*, **38**, 733–41.

Guiler, E.R. (1985). *Thylacine: the tragedy of the Tasmanian tiger*. Oxford University Press, Melbourne.

Guiler, E. & Godard, P. (1998). *Tasmanian Tiger: a lesson to be learnt*. Abrolhos Publishing, Perth.

Hamilton, A. (1972). Aboriginal man's best friend? *Mankind*, **8**, 287–95.

Harle, K.J. (1997). Late Quaternary vegetation and climate change in southeastern Australia: palynological evidence from marine core E55-6. *Palaeogeography, Palaeoclimatology, Palaeoecology*, **131**, 465–83.

Harle, K.J. *et al.* (2002). A chronology for the long pollen record from Lake Wangoom, western Victoria (Australia) as derived from uranium/thorium disequilibrium dating. *Journal of Quaternary Science*, **17**, 707–20.

Hayward, M. (2002). The ecology of the quokka (*Setonix brachyurus*) (Macropodidae: Marsupialia) in the northern jarrah forest of Australia. PhD thesis, University of New South Wales, Sydney.

Henderson, R.A. *et al.* (2000). Biogeographical observations on the Cretaceous biota of Australasia, in *Palaeobiogeography of Australasian Faunas and Floras* (eds A.J. Wright, G.C. Young, J.A. Talent & J.R. Laurie), pp. 355–404. Association of Australasian Palaeontologists, Canberra.

Hesse, P.P. (1994). The record of continental dust from Australia in Tasman Sea sediments. *Quaternary Science Reviews*, **13**, 257–72.

Hesse, P.P. & McTainsh, G.H. (2003). Australian dust deposits: modern processes and the Quaternary record. *Quaternary Science Reviews*, **22**, 2007–35.

Hesse, P.P., Magee, J.W. & van der Kaars, S. (2004). Late Quaternary climates of the Australian arid zone: a review. *Quaternary International*, **118–19**, 87–102.

Hill, R.S. (1994). The history of selected Australian taxa, in *History of the Australian vegetation: Cretaceous to recent* (ed. R.S. Hill), pp. 390–420. Cambridge University Press, Cambridge.

Hocknull, S.A. (2005). Ecological succession during the late Cainozoic of central eastern Queensland: extinction of a diverse rainforest community. *Memoirs of the Queensland Museum*, **51**, 39–122.

Holdaway, R.N. & Jacomb, C. (2000). Rapid extinction of the moas (Aves: Dinornithiformes): model, test and implications. *Science*, **287**, 2250-4.

Hope, J.H. (1978). Pleistocene mammal extinctions: the problem of Mungo and Menindee, New South Wales. *Alcheringa*, **2**, 65–82.

—— (1981). A new species of *Thylogale* (Marsupialia: Macropodidae) from Mapala Rock Shelter, Jaya (Carstensz) Mountains, Irian Jaya (Western New Guinea), Indonesia. *Records of the Australian Museum*, **33**, 369–87.

Hope, J.H., Dare-Edwards, A. & McIntyre, M.L. (1983). Middens and megafauna: stratigraphy and dating of Lake Tandou Lunette, Western New South Wales. *Archaeology in Oceania*, **18**, 38–45.

Hope, J.H. *et al.* (1977). Late Pleistocene faunal remains from Seton rock shelter, Kangaroo Island, South Australia. *Journal of Biogeography*, **4**, 363–85.

Horton, D. (1984). Red kangaroos: last of the Australian megafauna, in *Quaternary Extinctions: a prehistoric revolution* (eds P.S. Martin & R.G. Klein), pp. 639–80. University of Arizona Press, Tucson, Arizona.

Horton, D. (2000). *The Pure State of Nature*, Allen & Unwin, Sydney.

Horton, D.R. (1982). The burning question: Aborigines, fire and Australian ecosystems. *Mankind*, **13**, 237–51.

Hughes, P.J. & Lampert, R.J. (1982). Prehistoric population change in southern coastal New South Wales, in *Coastal Archaeology in Eastern Australia* (ed. S. Bowdler), pp. 16-29. Department of Prehistory, Research School of Pacific Studies, Australian National University, Canberra.

Hume, I.D. (1999). *Marsupial Nutrition*. Cambridge University Press, Cambridge.

Hurles, M.E., Matisoo-Smith, E., Gray, R.D. & Penny, D. (2003). Untangling Oceanic settlement: the edge of the knowable. *Trends in Ecology and Evolution*, **18**, 531–40.

Irlbeck, N.A. & Hume, I.D. (2003). The role of *Acacia* in the diets of Australian marsupials: a review. *Australian Mammalogy*, **25**, 121–34.

Jacobs, M. (1965). The genus *Capparis* (Capparaceae) from the Indus to the Pacific. *Blumea*, **12**, 385–541.

James, A. (2004). The creation of fertile patches by four ecosystem engineers in an arid South Australian dunefield. Hons thesis, University of New South Wales, Sydney.

Janzen, D.H. & Martin, P.S. (1982). Neotropical anachronisms: the fruits the gomphotheres ate. *Science*, **215**, 19-27.

Jarman, P.J. (1986). The red fox: an exotic, large predator, in *The Ecology of Exotic*

Animals and Plants: some case histories (ed. R.L. Kitching), pp. 44–61. Wiley, Brisbane.

Johnson, B.J. *et al.* (1999). 65,000 years of vegetation change in central Australia and the Australian summer monsoon. *Science*, **284**, 1150–2.

Johnson, C.N. (1994). Nutritional ecology of a mycophagous marsupial in relation to production of hypogeous fungi. *Ecology*, **75**, 2015–21.

—— (1995). Interactions between fire, ectomycorrhizal fungi and a mycophagous marsupial in eucalyptus forest. *Oecologia*, **104**, 467–75.

—— (1996). Interactions between mammals and ectomycorrhizal fungi. *Trends in Ecology and Evolution*, **11**, 503–7.

—— (1998). The evolutionary ecology of wombats, in *Wombats* (eds R. Wells & P. Pridmore), pp. 33–41. Surrey Beatty & Sons, Sydney.

—— (2002). Determinants of loss of mammal species during the late Quaternary 'megafauna' extinctions: life history and ecology, but not body size. *Proceedings of the Royal Society of London B*, **269**, 2221–7.

—— (2005a). The remaking of Australia's ecology. *Science*, **309**, 255–6.

—— (2005b). What can the data on late survival of Australian megafauna tell us about the cause of their extinction? *Quaternary Science Reviews*, **24**, 2167–72.

Johnson, C.N., Delean, S. & Balmford, A. (2002). Phylogeny and the selectivity of extinction in Australian marsupials. *Animal Conservation*, **5**, 135–42.

Johnson, C.N. & Prideaux, G.J. (2004). Extinctions of herbivorous mammals in Australia's late Pleistocene in relation to their feeding ecology: no evidence for environmental change as cause of extinction. *Austral Ecology*, **29**, 553–7.

Johnson, C. N., Vernes, K. and Payne, A. (2005). Demography in relation to population density in two herbivorous marsupials: testing for source-sink dynamics versus independent regulation of population size. *Oecologia*, **143**, 70–6.

Johnson, C.N. & Wroe, S. (2003). Causes of extinction of vertebrates during the Holocene of mainland Australia: arrival of the dingo, or human impact? *The Holocene*, **13**, 941–8.

Johnson, K.A., Burbidge, A.A. & McKenzie, N.L. (1989). Australian Macropodidae: status, causes of decline and future research and management, in *Kangaroos, Wallabies and Rat-kangaroos* (eds G. Grigg, P. Jarman & I. Hume), pp. 641–57. Surrey Beatty & Sons, Sydney.

Johnson, K.A., Gibson, D.F., Langford, D.G. & Cole, J.R. (1996). Recovery of the mala *Lagorchestes hirsutus*: a 30-year unfinished journey, in *Back from the Brink: refining the threatened species recovery process* (eds S. Stephens & S. Maxwell), pp. 155–61. Surrey Beatty & Sons, Sydney.

Johnson, K.A. & Jarman, P.J. (1976). Records of wild life as pests in the Armidale district, 1812–1975, in *Agriculture, Forestry and Wildlife: conflict or coexistence* (ed. P.J. Jarman). University of New England, Armidale.

Johnson, K.A. & Roff, A.D. (1982). The western quoll, *Dasyurus geoffroii* (Dasyuridae, Marsupialia) in the Northern Territory: historical records from venerable sources, in *Carnivorous Marsupials* (ed. M. Archer), vol. 1, pp. 221–6. Royal Zoological Society of New South Wales, Sydney.

Jones, F.W. (1923–25). *The Mammals of South Australia*. South Australian Government Printer, Adelaide.

Jones, M.E., Smith, G.C. & Jones, S.M. (2004). Is anti-predator behaviour in eastern quolls (*Dasyurus viverrinus*) effective against introduced predators? *Animal Conservation*, **7**, 155–60.

Jones, M.E. *et al.* (2003). Carnivore concerns: problems, issues and solutions for conserving Australasia's marsupial carnivores, in *Predators with pouches* (ed. M. Jones, C. Dickman & M. Archer), pp. 422–34. CSIRO Publishing, Canberra.

Jones, R. (1968). The geographical background to the arrival of man in Australia and Tasmania. *Archaeology and Physical Anthropology in Oceania*, **3**, 186–215.

—— (1969). Fire stick farming. *Australian Natural History*, **16**, 224–8.

—— (1970). Tasmanian Aborigines and dogs. *Mankind*, **7**, 256–71.

—— (1985). Archaeological conclusions, in *Archaeological Research in Kakadu National Park* (ed. R. Jones), pp. 291–8. Australian National Parks and Wildlife Service, Canberra.

—— (1994). Mindjongork: legacy of the firestick, in *Country in Flames: proceedings of the 1994 symposium on biodiversity and fire in North Australia* (ed. D.B. Rose). Biodiversity Unit, Department of the Environment, Sport and Territories / North Australia Research Unit, Australian National University, Canberra.

—— (1995). Tasmanian archaeology: establishing the sequences. *Annual Review of Anthropology*, **24**, 423–46.

Jones, R. & Johnson, I. (1985). Deaf Adder Gorge: Lindner Site, Nauwalabila I, in *Archaeological Research in Kakadu National Park* (ed. R. Jones), pp. 165–228. Australian National Parks and Wildlife Service, Canberra.

Kendrick, G.W. & Porter, J.K. (1973). Remains of a thylacine (Marsupialia: Dasyuroidea) and other fauna from caves in the Cape Range, Western Australia. *Journal of the Royal Society of Western Australia*, **56**.

Kerle, J.A. (2001). *Possums: the brushtails, ringtails and greater glider*. University of New South Wales Press, Sydney.

Kershaw, A.P. (1986). Climatic change and Aboriginal burning in north-east Australia during the last two glacial interglacial cycles. *Nature*, **322**, 47–9.

—— (1995). Environmental change in Greater Australia. *Antiquity*, **69**, 656–75.

Kershaw, A.P., Clark, J.S., Gill, A.M., & D'Costa, D.M. (2002). A history of fire in Australia, in *Flammable Australia: the fire regimes and biodiversity of a continent* (eds R.A. Bradstock, J.E. Williams & A.M. Gill), pp. 3–25. Cambridge University Press, Cambridge.

Kershaw, A.P., Mckenzie, G.M. & McMinn, A. (1993). A Quaternary vegetation history of northeastern Queensland from pollen analysis of ODP site 820. *Proceedings of the Ocean Drilling Program, Scientific Results*, **133**, 107–14.

Kershaw, A.P., Martin, H.A. & McEwan Mason, J.R.C. (1994). The neogene: a period of transition, in *History of the Australian Vegetation: Cretaceous to recent* (ed. R.S. Hill), pp. 299–327. Cambridge University Press, Cambridge.

Kershaw, A.P., van der Kaars, S. & Moss, P.T. (2003). Late Quaternary Milankovitch-scale climatic change and variability and its impact on monsoonal Australasia. *Marine Geology*, **201**, 81–95.

Kershaw, A.P. *et al.* (2000). Palaeobiogeography of the Quaternary of Australia, in *Palaeobiogeography of Australasian Faunas and Floras* (eds A.J. Wright, G.C. Young, J.A. Talent & J.R. Laurie), pp. 471–515. Association of Australasian Palaeontologists, Canberra.

Kinnear, J.E., Onus, M.L. & Sunmner, N.R. (1998). Fox control and rock-wallaby population dynamics, II: an update. *Wildlife Research*, **25**, 81–8.

Kinnear, J.E., Sumner, N.R. & Onus, M.L. (2002). The red fox in Australia: an exotic predator turned biocontrol agent. *Biological Conservation*, **108**, 335–59.

Kohen, J.L. (1995). *Aboriginal Environmental Impacts*. University of New South Wales Press, Sydney.

Krajewski, C. & Westerman, M. (2003). Molecular systematics of Dasyuromorphia, in *Predators with Pouches: the biology of carnivorous marsupials* (eds M.E. Jones, C.R. Dickman & M. Archer), pp. 3–20. CSIRO Publishing, Melbourne.

Krefft, G. (1866). On the vertebrate animals of the Lower Murray and Darling, their habits, economy, and geographical distribution. *Transactions of the Philosophical Society of New South Wales*, **1862–65**, 1–33.

Lambeck, K. & Chappell, J. (2001). Sea level change through the last glacial cycle. *Science*, **292**, 679–385.

Lambeck, K., Yokoyama, Y., & Purcell, T. (2002). Into and out of the Last Glacial Maximum: sea-level change during Oxygen Isotope Stages 3 and 2. *Quaternary Science Reviews*, **21**, 343–60.

Law, R. (2001). Phenotypic and genetic change due to selective exploitation, in *Conservation of Exploited Species* (eds J.D. Reynolds, G.M. Mace, K.H. Redford & J.G. Robinson), pp. 323–42. Cambridge University Press, Cambridge.

Le Souef, A.S. & Burrell, H. (1926). *The Wild Animals of Australasia*. George G. Harrap & Co., London.

Leavesley, M. & Allen, J. (1998). Dates, disturbance and artefact distributions: another analysis of Buang Merabak, a Pleistocene site on New Ireland, Papua New Guinea. *Archaeology in Oceania*, **33**, 63–82.

Leavesley, M.G. *et al.* (2002). Buang Merabak: early evidence for human occupation in the Bismarck archipelago, Papua New Guinea. *Australian Archaeology*, **54**, 55–7.

Lee, A.K. (1995). *The Action Plan for Australian Rodents*. Australian Nature Conservation Agency, Canberra.

Leigh, J.H. (1994). Chenopod shrublands, in *Australian Vegetation* (ed. R.H. Groves), pp. 345–67. Cambridge University Press, Cambridge.

Leonard, J.A. *et al.* (2002). Ancient DNA evidence for old world origin of new world dogs. *Science*, **298**, 1613–16.

Lewis, D. (1986). 'The dreamtime animals': a reply. *Archaeology in Oceania*, **21**, 140–14.

Lillegraven, J.A. (1974). Biogeographical considerations of the marsupial–placental dichotomy. *Annual Review of Ecology and Systematics*, **5**, 263–83.

Long, J., Archer, M., Flannery, T. & Hand, S. (2002). *Prehistoric Mammals of Australia and New Guinea*. University of New South Wales Press, Sydney.

Long, J.L. (2003). *Introduced Mammals of the World*. CSIRO Publishing, Melbourne.

Longmore, M.E. & Heijnis, H. (1999). Aridity in Australia: Pleistocene records of palaeohydrological and palaeoecological change from the perched lake sediments of Fraser Island, Queensland, Australia. *Quaternary International*, **57/58**, 35–47.

Lourandos, H. (1983). Intensification: a Late Pleistocene–Holocene archaeological sequence from Southwestern Victoria. *Archaeology in Oceania*, **18**, 81–94.

Lourandos, H. (1997). *Continent of Hunter-gatherers: new perspectives in Australian prehistory*. Cambridge University Press, Cambridge.

Lourandos, H. & David, B. (2002). Long-term archaeological and environmental trends: a comparison from Late Pleistocene–Holocene Australia, in *Bridging Wallace's Line: the environmental and cultural history and dynamics of the southeast Asian-Australasian region* (eds A.P. Kershaw, N.J. Tapper, B. David, P. Bishop & D. Penny), pp. 307–38. Catena Verlag, Cremlingen, Germany.

Low, T. (1998). Thorny thoughts. *Nature Australia*, **summer 1997–98**, 22–3.

Lowry, J.W.J. & Merrilees, D. (1969). Age of the desiccated carcase of a thylacine (Marsupialia, Dasyuroidea) from thylacine hole, Nullarbor region, Western Australia. *Helictite*, **7**, 15–16.

Luebbers, R.A. (1975). Ancient boomerangs discovered in South Australia. *Nature*, **253**, 39.

Luly, J. (2001a). Palynology of the perfumed pineries of arid South Australia, in *Perfumed Pineries: environmental history of Australia's* Callitris *forests* (eds J. Dargavel, D. Hart & B. Libbis). Centre for Resource and Environmental Studies, Australian National University, Canberra.

—— (2001b). On the equivocal fate of Late Pleistocene *Callitris* Vent. (Cupressaceae) woodlands in arid South Australia. *Quaternary International*, **83–85**, 155–68.

Lunney, D. (2001). Causes of the extinction on native mammals of the Western Division of New South Wales: an ecological interpretation of the nineteenth century historical record. *Rangeland Journal*, **23**, 44–70.

Lunney, D. & Leary, D. (1988). The impact of native mammals of land-use changes and exotic species in the Bega district, New South Wales, since settlement. *Australian Journal of Ecology*, **13**, 67–92.

Luo, Z.-X., Ji, Q., Wible, J.R. & Yuan, C.-X. (2003). An early Cretaceous tribosphenic mammal and metatherian evolution. *Science*, **302**, 1934–40.

Lyons, S.K. *et al.* (2004). Was a 'hyperdisease' responsible for the late Pleistocene megafauna extinction? *Ecology Letters*, **7**, 859–68.

McBryde, I. (1982). *Coast and estuary: archaeological investigations on the north coast of New South Wales at Wombah and Schnapper Point*. Australian Institute of Aboriginal Studies, Canberra.

McGowran, B. *et al.* (2000). Australasian paleobiogeography: the paleogene and neogene record, in *Palaeobiogeography of Australasian Faunas and Floras* (eds A.J. Wright, G.C. Young, J.A. Talent & J.R. Laurie), pp. 405–70. Association of Australian Palaeontologists, Canberra.

McIlwee, A.P. & Johnson, C.N. (1998). Nutritional value of fungus to three marsupial herbivores, revealed by stable isotope analysis. *Functional Ecology*, **12**, 223–31.

Mckenzie, G.M. & Kershaw, A.P. (2000). The last glacial cycle from Wyelangta, the Otway region of Victoria, Australia. *Palaeogeography, Palaeoclimatology, Palaeoecology*, **155**, 177–93.

McKenzie, N.L. *et al.* (2005). Regional patterns in the attrition of Australia's mammal fauna (abstract). *Newsletter of the Australian Mammal Society*, **Oct 2005**.

McNamara, J.A. (1997). Some smaller macropod fossils of South Australia. *Proceedings of the Linnean Society of New South Wales*, **117**, 97–105.

MacPhee, R.D.E. & Flemming (1999). *Requiem Aeternum*: the last five hundred years of mammalian species extinctions, in *Extinctions in near time: causes, contexts, and consequences* (ed. R.D.E. MacPhee), pp. 333–72. Kluwer Academic/Plenum Publishers, New York.

MacPhee, R.D.E. & Marx, P.A. (1997). The 40,000-year plague: humans, hyperdisease, and first-contact extinctions, in *Natural Change and Human Impact in Madagascar* (eds S. Goodman & B. Patterson), pp. 169–217. Smithsonian Institution Press, Washington, DC.

Magee, J.W., Bowler, J.M., Miller, G.H. & Williams, D.L.G. (1995). Stratigraphy, sedimentology, chronology and palaeohydrology of Quaternary lacustrine deposits at Madigan Gulf, Lake Eyre, South Australia. *Palaeogeography, Palaeoclimatology, Palaeoecology*, **113**, 3–42.

Main, A.R. (1978). Ecophysiology: towards an understanding of Late Pleistocene marsupial extinction, in *Biology and Quaternary Environments* (eds D. Walker & J.C. Guppy), pp. 169–84. Australian Academy of Science, Canberra.

Marsack, P. & Campbell, G. (1990). Feeding behaviour and diet of dingoes in the Nullarbor region of Western Australia. *Australian Wildlife Research*, **17**, 349–58.

Marshall, L.G. & Corruccini, R.S. (1978). Variability, evolutionary rates and allometry in dwarfing lineages. *Paleobiology*, **4**, 101–19.

Martin, H.A. (1990). Tertiary climate and phytogeography in southeastern Australia. *Review of Palaeobotany and Palynology*, **65**, 47–55.

Martin, P.S. (1973). The discovery of America. *Science*, **179**, 969–74.

—— (1984). Prehistoric overkill: the global model, in *Quaternary Extinctions* (eds P.S. Martin & R.G. Klein), pp. 354–403. University of Arizona Press, Tucson, Arizona.

Martin, P.S. & Steadman, D.W. (1999). Prehistoric extinctions on islands and continents, in *Extinctions in Near Time: causes, contexts and consequences* (ed. R.D.E. MacPhee), pp. 17–56. Kluwer Academic / Plenum Publishers, New York.

Martin, R. (1996). Tcharibeena: field studies of Bennett's tree-kangaroo, in *Tree Kangaroos: a curious natural history* (eds T.F. Flannery, R. Martin & A. Szalay), pp. 36–65. Reed Books, Melbourne.

Martin, R. & Handasyde, K. (1999). *The Koala: natural history, conservation and management*. University of New South Wales Press, Sydney.

Martinez-Meyer, E., Peterson, A.T. & Hargrove, W.W. (2004). Ecological niches as stable distributional constraints on mammal species, with implications for Pleistocene extinctions and climate change projections for biodiversity. *Global Ecology and Biogeography*, 13, 305–14.

Maxwell, S., Burbidge, A.A. & Morris, K., eds. (1996). *The 1996 Action Plan for Australian Marsupials and Monotremes*. Environment Australia, Canberra.

Meggitt, M.J. (1965). The association between Australian Aborigines and dingoes, in *Man, Culture and Animals* (eds A. Leeds & A.P. Vayda), pp. 7–26. American Association for the Advancement of Science, Washington, DC.

Menkhorst, P.W. (1995). *Mammals of Victoria*. Oxford University Press, Melbourne.

Merrilees, D. (1968). Man the destroyer: late Quaternary changes in the Australian marsupial fauna. *Journal of the Royal Society of Western Australia*, **51**, 1–24.

Milham, P. & Thompson, P. (1976). Relative antiquity of human occupation and extinct fauna at Madura Cave, southeastern Western Australia. *Mankind*, **10**, 175–80.

Miller, G.H., Magee, J.W. & Jull, A.J.T. (1997). Low-latitude glacial cooling in the southern hemisphere from amino-acid racemization in emu eggshells. *Nature*, **385**, 241–4.

Miller, G.H. *et al.* (1999). Pleistocene extinction of *Genyornis newtoni*: human impact on Australian megafauna. *Science*, 283, 205–8.

Miller, G.H. *et al.* (2005). Ecosystem collapse in Pleistocene Australia and a human role in megafaunal extinction. *Science*, **309**, 287–90.

Mitchell, B.D. & Banks, P.B. (2005). Do wild dogs exclude foxes? Evidence for competition from dietary and spatial overlaps. *Austral Ecology*, **30**, 581–91.

Mix, A.C. *et al.* (1995). Benthic foraminifer stable isotope record from Site 849 (0–5 MA): local and global climate changes. *Proceedings of the Ocean Drilling Program, Scientific Results*, **138**, 371–412.

Molnar, R.E. (1985). *Quinkana fortirostrum* The Quinkan, in *Kadimakara: extinct vertebrates of Australia* (eds P.V. Rich & G.F. van Tets), pp. 160–5. Pioneer Design Studio, Melbourne.

Molnar, R.E. & Kurz, C. (1997). The distribution of Pleistocene vertebrates on the eastern Darling Downs, based on the Queensland Museum collections. *Proceedings of the Linnean Society of New South Wales*, **117**, 107–34.

Moriarty, K.C., McCulloch, M.T., Wells, R.T., & McDowell, M.C. (2000). Mid-Pleistocene cave fills, megafaunal remains and climate change at Naracoorte, South Australia: towards a predictive model using U-Th dating of speleothems. *Palaeogeography, Palaeoclimatology, Palaeoecology*, **159**, 113–43.

Morris, K. *et al.* (2003). Recovery of the threatened chuditch (*Dasyurus geoffroii*): a case study. In *Predators with Pouches: the biology of carnivorous marsupials* (eds M.E. Jones, C. Dickman & M. Archer), pp. 435–51. CSIRO Publishing, Melbourne.

Morse, K. (1993). Who can see the sea?: prehistoric Aboriginal occupation of the Cape Range peninsula. *Records of the Western Australian Museum*, **supplement 45**, 227–42.

Morton, S.R. (1990a). The impact of European settlement on the vertebrate animals of arid Australia: a conceptual model, in *Australian ecosystems: 200 years of utilization, degradation and reconstruction* (eds D.A. Saunders, A.J.M. Hopkins & R.A. How), pp. 201–13. Ecological Society of Australia,

—— (1990b). Determinants of diversity in animal communities of arid Australia, in *Species Diversity in Ecological Communities* (eds R.E. Ricklefs & D. Schulter), pp. 159–69. University of Chicago Press, Chicago.

Morwood, M.J. (1987). The archaeology of social complexity in south-east Queensland. *Proceedings of the Prehistoric Society*, **53**, 337–50.

—— (2002). *Visions from the Past: the archaeology of Australian Aboriginal art.* Allen & Unwin, Sydney.

Morwood, M.J. & Hobbs, D.R. (1995). Themes in the prehistory of tropical Australia. *Antiquity*, **69**, 747–68.

Morwood, M.J. & Trezise, P.J. (1989). Edge-ground axes in Pleistocene Greater Australia: new evidence from S.E. Cape York Peninsula. *Queensland Archaeological Research*, **6**, 77–90.

Moss, P. & Kershaw, A.P. (2000). The last glacial cycle from the humid tropics of northeastern Australia: comparison of a terrestrial and a marine record. *Palaeogeography, Palaeoclimatology, Palaeoecology*, **155**, 155–76.

Muizon, C.D., Cifelli, R.L. & Paz, R.C. (1997). The origin of the dog-like borhyaenoid marsupials of South America. *Nature*, **389**, 486–9.

Mulvaney, J. & Kamminga, J. (1999). *Prehistory of Australia.* Allen & Unwin, Sydney.

Mulvaney, D.J. *et al.* (1964). Archaeological excavation of rock shelter no. 6 Fromm's Landing, South Australia. *Proceedings of the Royal Society of Victoria*, **77**, 479–516.

Munyikwa, K. (2005). Synchrony of southern hemisphere late Pleistocene arid episodes: a review of luminescence chronologies from arid aeolian landscapes south of the equator. *Quaternary Science Reviews*, **24**, 2555–83.

Murphy, M.T., Garkaklis, M.J. & Hardy, G.E.S.J. Seed caching by woylies *Bettongia penicillata* can increase sandalwood *Santalum spicatum* regeneration in Western Australia. *Austral Ecology*, **in press**.

Murray, P. (1984). Extinctions downunder: a bestiary of extinct Australian Late Pleistocene monotremes and marsupials, in *Quaternary extinctions: a prehistoric revolution* (eds P.S. Martin & R.G. Klein), pp. 600–29. University of Arizona Press, Tucson, Arizona.

Murray, P. (1991). The Pleistocene megafauna of Australia, in *Vertebrate palaeontology of Australasia* (eds P. Vickers-Rich, J.M. Monaghan, R.F. Baird & T.H. Rich), pp. 1071–64. Pioneer Design Studio/Monash University, Melbourne.

Murray, P. & Chaloupka, G. (1984). The dreamtime animals: extinct megafauna in Arnhem Land rock art. *Archaeology in Oceania*, **19**, 105–16.

Murray, P.F. (1992). Thinheads, thickheads and airheads: functional craniology of some diprotodontian marsupials. *The Beagle: Records of the Northern Territory Museum of Arts and Sciences*, **9**, 71–88.

Murray, P.F. (1998). Palaeontology and palaeobiology of wombats, in *Wombats* (eds R.T. Wells & P.A. Pridmore), pp. 1–33. Surrey Beatty & Sons, Sydney.

Murray, P.F. & Vickers-Rich, P. (2004). *Magnificent Mihirungs: the colossal flightless birds of the Australian dreamtime*. Indiana University Press, Bloomington and Indianapolis, Indiana.

Musser, A.M. (1999). Diversity and relationships of living and extinct monotremes. *Australian Mammalogy*, **21**, 8–9.

—— (2003). Review of the monotreme fossil record and comparison of palaeontological and molecular data. *Comparative Biochemistry and Physiology, part A*, 136, 927-942.

Nanson, G.C., Price, D.M. & Short, S.A. (1992). Wetting and drying of Australia over the past 300 ka. *Geology*, **20**, 791-794.

Newsome, A.E. (1975). An ecological comparison of the two arid-zone kangaroos of Australia, and their anomalous prosperity since the introduction of ruminant stock to their environment. *Quarterly Review of Biology*, **50**, 389-424.

—— (2001). The biology and ecology of the dingo, in *A Symposium on the Dingo* (eds C.R. Dickman & D. Lunney), pp. 20-33. Royal Zoological Society of New South Wales, Sydney.

Newsome, A.E., Catling, P.C., Cooke, B.D. & Smyth, R. (2001). Two ecological universes separated by the dingo barrier fence in semi-arid Australia: interaction between landscapes, herbivory and carnivory, with and without dingoes. *Rangeland Journal*, **23**, 71-98.

Nolche, G. (2001). Where is the smoking gun? *Australasian Science*, **Sept**, 19-20.

Oakwood, M. (2000) Reproduction and demography of the northern quoll, *Dasyurus hallucatus*, in the lowland savanna of northern Australia. *Australian Journal of Zoology*, **48**, 519-539.

O'Connell, J.F. & Allen, J. (2004). Dating the colonization of Sahul (Pleistocene Australia – New Guinea): a review of recent research. *Journal of Archaeological Science*, **31**, 835-853.

O'Connor, S. (1999). *30,000 Years of Aboriginal Occupation: Kimberley, North west Australia* ANH Publications/The Centre for Archaeological Research, Australian National University, Canberra.

O'Connor, S. *et al.* (2002). From savanna to rainforest: changing environments and human occupation at Liang Lembudu, Aru Islands, Maluku (Indonesia), in *Bridging Wallace's Line: the environmental and cultural history and dynamics of the southeast Asian–Australasian region* (eds A.P. Kershaw, N.J. Tapper, B. David, P. Bishop & D. Penny), pp. 279-306. Catena Verlag, Cremlingen, Germany.

O'Connor, S., Veth, P. & Hubbard, N. (1993). Changing interpretations of post-glacial human subsistence and demography in Sahul, in *Sahul in Review: Pleistocene archaeology in Australia, New Guinea and Island Melanesia* (eds M.A. Smith, M. Spriggs & B. Frankhauser), pp. 95-105. Department of Prehistory, Research School of Pacific Studies, Australian National University, Canberra.

Owen, R. (1861). *Palaeontology, or, A Systematic Study of Extinct Animals and Their Geological Relations*, second edition. Adam and Charles Black, Edinburgh.

—— (1877). *Researches on the Fossil Remains of the Extinct Mammals of Australia; with a notice of the extinct marsupials of England*. Erxleben, London.

Owen-Smith, N. (1999). The interaction of humans, megaherbivores and habitats in the late Pleistocene extinction event, in *Extinctions in Near Time: causes, contexts and consequences* (ed. R.D.E. MacPhee), pp. 57–70. Kluwer Academic/ Plenum Publishers, New York.

Owen-Smith, R.N. (1988). *Megaherbivores: the influence of very large body size on ecology*. Cambridge University Press, Cambridge.

Pack, S.M., Miller, G.H., Fogel, M.L. & Spooner, N.A. (2003). Carbon isotope evidence for increased aridity in northwestern Australia through the Quaternary. *Quaternary Science Reviews*, **22**, 629–43.

Paddle, R.N. (2000). *The Last Tasmanian Tiger*. Cambridge University Press, Cambridge.

Palomares, F. & Caro, T.M. (1999). Interspecific killing among mammalian carnivores. *American Naturalist*, **153**, 492–508.

Parris, H.S. (1948). Koalas on the lower Goulburn. *Victorian Naturalist*, **64**, 192–3.

Partridge, J. (1967). A 3,300 year old thylacine (Marsupialia: Thylacinidae) from the Nullarbor Plain, Western Australia. *Journal of the Royal Society of Western Australia*, **50**, 57–9.

Pate, F.D., McDowell, M.C., Wells, R.T. & Smith, A.M. (2002). Last recorded evidence for megafauna at Wet Cave, Naracoorte, South Australia 45,000 years ago. *Australian Archaeology*, **54**, 53–5.

Pavlides, C. & Gosden, C. (1994). 35,000-year-old sites in the rainforests of West New Britain, Papua New Guinea. *Antiquity*, **68**, 604–10.

Pearson, S. *et al.* (1999). The spatial and temporal patterns of stick-nest rat middens in Australia. *Radiocarbon*, **41**, 295–308.

Petit, J.R. *et al.* (1999). Climate and atmospheric history of the past 420 000 years from the Vostok ice core, Antarctica. *Nature*, **399**, 429-36.

Plane, M.D. (1976). The occurrence of *Thylacinus* in Tertiary rocks from Papua New Guinea. *Journal of Australian Geology and Geophysics*, **1**, 78–9.

Pledge, N.S., Prescott, J.R. & Hutton, J.T. (2002). A late Pleistocene occurrence of *Diprotodon* at Hallett Cove, South Australia. *Transactions of the Royal Society of South Australia*, **126**, 39–44.

Plomley, N.J.B. (1966). *Friendly Mission: the Tasmanian journals and papers of George Augustus Robinson 1829–1834*. Tasmanian Historical Research Association, Hobart.

Price, G.J. & Hocknull, S.A. (2005a). A small adult *Palorchestes* (Marsupialia, Palorchestidae) from the Pleistocene of the Darling Downs, southeast Queensland. *Memoirs of the Queensland Museum*, **51**, 202.

Price, O. *et al.* (2005b). Regional patterns of mammal abundance and their relationship to landscape variables in eucalypt woodlands near Darwin, northern Australia. *Wildlife Research*, **32**, 435–46.

Prideaux, G.J. (1999). *Borungaboodie hatcheri* gen. et sp. nov., a very large bettong (Marsupialia: Macropodoidea) from the Pleistocene of southwestern Australia. *Records of the Western Australian Museum*, **supplement 57**, 317–29.

—— (2004). *Systematics and evolution of the sthenurine kangaroos*. University of California Publications in Geological Sciences, vol. 146. University of California Press, Berkeley.

Prideaux, G.J. *et al.* (2000). Tight Entrance Cave, southwestern Australia: a late Pleistocene vertebrate deposit spanning more than 180 ka. *Journal of Vertebrate Paleontology*, **20** (suppl. to no. 3), 62A.

Purvis, A. (2001). Mammalian life histories and responses to population exploitation, in *Conservation of Exploited Species* (eds J.D. Reynolds, G.M. Mace, K.H. Redford & J.G. Robinson), pp. 169–81. Cambridge University Press, Cambridge.

Quilty, P.G. (1994). The background: 144 million years of Australian paleoclimate and paleogeography, in *History of the Australian Vegetation: Cretaceous to Recent* (ed. R.S. Hill). Cambridge University Press, Cambridge.

Ratcliffe, F. (1947). *Flying Fox and Drifting Sand*. Angus & Robertson, Sydney.

Read, J. & Bowen, Z. (2001). Population dynamics, diet and aspects of the biology of feral cats and foxes in arid South Australia. *Wildlife Research*, **28**, 195–201.

Reed, E.H. & Bourne, S.J. (2000). Pleistocene fossil vertebrate sites of the south east region of South Australia. *Transactions of the Royal Society of South Australia*, **124**, 61–90.

Rich, T.H. (1985). *Megalania prisca*: the giant goanna, in *Kadimakara: extinct vertebrates of Australia* (eds P.V. Rich & G.F. van Tets), pp. 152–5. Pioneer Design Studio, Melbourne.

Rich, T.H. et al. (1997). A tribosphenic mammal from the Mesozoic of Australia. *Science*, **278**, 1438–42.

—— (1999). Early Cretaceous mammals from Flat Rocks, Victoria, Australia. *Records of the Queen Victoria Museum*, **106**, 1–34.

Ride, W.D.L. et al. (1997). Towards a biology of *Propleopus oscillans* (Marsupialia: Propleopinae, Hypsiprymnodontidae). *Proceedings of the Linnean Society of New South Wales*, **117**, 243–328.

Risbey, D.A. *et al.* (2000). The impact of cats and foxes on the small vertebrate fauna of Herisson Prong, Western Australia: II. A field experiment. *Wildlife Research*, **27**, 223–35.

Roberts, R.G. *et al.* (2001a). The last Australian megafauna. *Australasian Science*, **22**, 40–1.

—— (2001b). New ages for the last Australian megafauna: continent-wide extinction about 46,000 years ago. *Science*, **292**, 1888–92.

Roberts, R.G., Jones, R. & Smith, M.A. (1990a). Early dates at Malakunanja II: a reply to Bowdler. *Australian Archaeology*, **31**, 94–7.

—— (1990b). Thermoluminescence dating of a 50,000-year-old human occupation site in northern Australia. *Nature*, **345**, 153–6.

Roberts, R.G. (1996). Preliminary luminescence dates for archaeological sediments on the Nullarbor Plain, South Australia. *Australian Archaeology*, **42**, 7–16.

Roberts, R.G. *et al.* (1994). The human colonization of Australia: optical dates of 53,000 and 60,000 years bracket human arrival at Deaf Adder Gorge, Northern Territory. *Quaternary Science Reviews*, **13**, 573–84.

Robinson, A.C. & Young, M.C. (1983). The toolache wallaby, Rep. No. 2. South Australian Department of Environment and Planning, Adelaide.

Rolls, E.C. (1969). *They All Ran Wild* Angus & Robertson, Sydney.

Ross, A. (1981). Holocene environment and prehistoric site patterning in the Victorian Mallee. *Archaeology in Oceania*, **16**, 145–54.

—— (1985). Archaeological evidence for population change in the middle to late Holocene in southeastern Australia. *Archaeology in Oceania*, **20**, 81–9.

Sage, R.F., Li, M. & Monson, R.K. (1999). The taxonomic distribution of C4 photosynthesis, in *C4 Plant Biology* (eds R.F. Sage & R.K. Monson), pp. 551–84. Academic Press, San Diego, California.

Sanson, G.D. (1982). Evolution of feeding adaptations in fossil and recent macropodids. In *The Fossil Vertebrate Record of Australasia* (eds P.V. Rich & E.M. Thompson), pp. 489–506. Monash University, Melbourne.

Sanson, G.D., Riley, S.J. & Williams, M.A. (1980). A late Quaternary *Procoptodon* fossil from Lake George, New South Wales. *Search*, **11**, 39–40.

Saunders, G., Coman, B., Kinnear, J. & Braysher, M. (1995). *Managing Vertebrate Pests: foxes*. Australian Government Publishing Service, Canberra.

Savolainen, P. *et al.* (2002). Genetic evidence for an east Asian origin of domestic dogs. *Science*, **298**, 1610–13.

—— (2004). A detailed picture of the origin of the Australian dingo, obtained from the study of mitochondrial DNA. *Proceedings of the National Academy of Science of the USA*, 101, 12 387–90.

Schmitz, A. (2005). Summary of mammal translocations undertaken by the Australian Wildlife Conservancy (abstract). *Newsletter of the Australian Mammal Society*, **Oct 2005**.

Sepkoski, J.J. (1998). Rates of speciation in the fossil record. *Proceedings of the Royal Society of London B*, **353**, 315–26.

Short, J. (1998). The extinction of rat-kangaroos (Marsupialia: Potoroidae) in New South Wales, Australia. *Biological Conservation*, **86**, 365–77.

Short, J. (2004). Mammal declines in southern Western Australia: perspectives from Shortridge's collections of mammals in 1904–07. *Australian Zoologist*, **32**, 605–28.

Short, J. & Calaby, J.H. (2001). The status of Australian mammals in 1922: collections and field notes of museum collector Charles Hoy. *Australian Zoologist*, **31**, 533–62.

Short, J., Kinnear, J.E. & Robley, A. (2002). Surplus killing by introduced predators in Australia: evidence for ineffective anti-predator adaptations in native prey species? *Biological Conservation*, **103**, 283–301.

Short, J. & Turner, B. (1994). A test of the vegetation mosaic hypothesis to explain the decline and extinction of Australian mammals. *Conservation Biology*, **8**, 439–49.

Short, J. *et al.* (1992). Reintroductions of macropods (Marsupialia: Macropodidae) in Australia: a review. *Biological Conservation*, **62**, 189–204.

Siegenthaler, U. *et al.* (2005). Stable carbon cycle–climate relationship during the late Pleistocene. *Science*, **310**, 1313–17

Sinclair, A.R.E. *et al.* (1998). Predicting effects of predators on conservation of endangered prey. *Conservation Biology*, **12**, 564–75.

Sinclair, E.A., Danks, A. & Wayne, A.F. (1996). Rediscovery of Gilbert's potoroo, *Potorous tridactylus*, in Western Australia. *Australian Mammalogy*, **19**, 69–72.

Singh, G. & Geissler, E.A. (1985). Late Cainozoic history of vegetation, fire, lake levels and climate at Lake George, New South Wales, Australia. *Philosophical Transactions of the Royal Society of London B*, **311**, 379–447.

Singh, G., Kershaw, A.P. & Clarke, R. (1981). Quaternary vegetation and fire history in Australia, in *Fire and the Australian Biota* (eds A.M. Gill, R.H. Groves & I.R. Noble), pp. 23–54. Australian Academy of Sciences, Canberra.

Slack, M.J., Fullagar, R.L.K., Field, J.H. & Border, A. (2004). New Pleistocene ages for backed artefact technology in Australia. *Archaeology in Oceania*, **39**, 131–7.

Smith, A.G., Smith, D.G. & Funnell, B.M. (1994). *Atlas of Mesozoic and Cenozoic Coastlines*. Cambridge University Press, Cambridge.

Smith, A.P. & Quin, D.G. (1996). Patterns and causes of extinction and decline in Australian conilurine rodents. *Biological Conservation*, **77**, 243–67.

Smith, M.A. (1982). Devon Downs reconsidered: changes in site use at a lower Murray Valley rockshelter. *Archaeology in Oceania*, **17**, 109–16.

—— (2005). Paleoclimates: an archaeology of climate change, in *A Change in the Weather: climate and culture in Australia* (eds T. Sherratt, T. Griffiths & L. Robin), pp. 176–86. National Museum of Australia Press, Canberra.

Smith, M.A. & Sharp, N.D. (1993). A revised bibliography of Pleistocene archaeological sites in Australia, New Guinea and island Melanesia, in *Sahul in Review* (eds M.A. Smith, M. Spriggs & B. Frankhauser), pp. 283–312. Department of Prehistory, Research School of Pacific Studies, Australian National University, Canberra.

Smith, M.A. *et al.* (2001). New ABOX AMS-14C ages remove dating anomalies at Puritjarra rock shelter. *Australian Archaeology*, **53**, 45–7.

Southgate, R.I. (1990). Distribution and abundance of the greater bilby *Macrotis lagotis* Reid (Marsupialia: Peramelidae), in *Bandicoots and Bilbies* (eds J.H. Seebeck, P.R. Brown, R.L. Wallis & C.M. Kemper), pp. 293–302. Surrey Beatty & Sons, Sydney.

Steadman, D.W. (1995). Prehistoric extinction of Pacific Island birds: biodiversity meets zooarcheology. *Science*, **267**, 1123–31.

Stirling, E.C. (1900). The physical features of Lake Callabonna. *Memoirs of the Royal Society of South Australia*, **1**, I–xv.

Strahan, R. (1995). *The Mammals of Australia*. Reed Books, Sydney.

Strahan, R. & Martin, R. (1982). The koala: little fact, much emotion. In *Species at risk: research in Australia* (eds R.H. Groves & W.D.L. Ride), pp. 147–58. Australian Academy of Science, Canberra.

Symon, D.E. (1981). A revision of the genus *Solanum* in Australia. *Journal of the Adelaide Botanic Gardens*, **4**, 1–367.

—— (1986). A survey of *Solanum* prickles and marsupial herbivory in Australia. *Annals of the Missouri Botanical Garden*, **73**, 745–54.

Szalay, F.S. (1994). *Evolutionary History of the Marsupials and an Analysis of Osteological Characters*. Cambridge University Press, Cambridge.

Tacon, P.S.C. & Brockwell, S. (1995). Arnhem Land prehistory in landscape, stone and paint. *Antiquity*, **69**, 676–95.

Tacon, P.S.C. & Chippindale, C. (1994). Australia's ancient warriors: changing depictions of fighting in the rock art of Arnhem Land, N.T. *Cambridge Archaeological Journal*, **4**, 211–48.

—— (2002). Changing places: north Australian rock-art transformations 4000–6000 BP, in *Ninth International Conference on Hunting and Gathering Societies*. Published online: http://www.abdn.ac.uk/chags9/1tacon.htm, Edinburgh, Scotland.

Taylor, R.H. (1979). How the Macquarie Island parakeeet became extinct. *New Zealand Journal of Ecology*, **2**, 42-5.

Tedford, R. (1967). The fossil Macropodidae from Lake Menindee, New South Wales. *University of California Publications in Geologial Science*, **64**, 1-156.

Tedford, R.H. (1955). Report on the extinct mammalian remains at Lake Menindee, New South Wales. *Records of the South Australian Museum*, **11**, 299-305.

—— (1973). The diprotodons of Lake Callabonna. *Australian Natural History*, **17**, 349-54.

—— (1984). The diprotodons of Callabonna, in *Vertebrate Zoogeography and Evolution in Australia* (eds M. Archer & G. Clayton), pp. 999-1002. Hesperian Press, Perth.

Thorley, P.B. (1998). Pleistocene settlement in the Australian arid zone: occupation of an inland riverine landscape in the central Australian ranges. *Antiquity*, **72**, 34-45.

Thorne, A. *et al.* (1999). Australia's oldest human remains: age of the Lake Mungo 3 skeleton. *Journal of Human Evolution*, **36**, 591–612.

Tindale, N. (1959). Ecology of primitive aboriginal man in Australia, in *Biogeography and Ecology in Australia* (eds A. Keast, R.L. Crocker & C.S. Christian), pp. 36–51. Junk, The Haag.

—— (1964). Radiocarbon dates of interest to Australian archaeologists. *Australian Journal of Science*, **27**, 24.

Torgersen, T. *et al.* (1988). Late Quaternary environments of the Carpentaria basin, Australia. *Palaeogeography, Palaeoclimatology, Palaeoecology*, **67**, 245–61.

Trueman, C.N.G. *et al.* (2005). Prolonged coexistence of humans and megafauna in Pleistocene Australia. *Proceedings of the National Academy of Science of the USA*, **102**, 8381–5.

Tuck, G.N., Polacheck, T., Croxall, J.P. & Weimerskirch, H. (2001). Modelling the impact of fishery by-catches on albatross populations. *Journal of Applied Ecology*, **28**, 1182–96.

Tudhope, A.W. *et al.* (2001). Variability in the El Niño–Southern Oscillation through a glacial-interglacial cycle. *Science*, **291**, 1511–17.

Tunbridge, D. (1991). *The Story of the Flinders Ranges Mammals*. Kangaroo Press, Sydney.

Turney, C.S.M. & Bird, M.I. (2002). Determining the timing and pattern of human colonisation in Australia: proposals for radiocarbon dating 'early' sequences. *Australian Archaeology*, **54**, 1–5.

Turney, C.S.M. *et al.* (2001a). Early human occupation at Devil's Lair, southwestern Australia 50,000 years ago. *Quaternary Research*, **55**, 3–13.

—— (2001b). Redating the onset of burning at Lynch's Crater (North Queensland): implications for human settlement in Australia. *Journal of Quaternary Science*, **16**, 767–71.

Tyndale-Biscoe, H. (2005). *Life of marsupials*. CSIRO Publishing, Melbourne.

van der Kaars, S. & De Deckker, P. (2002). A late Quaternary pollen record from deep-sea core Fr10/95 GC17 offshore Cape Range Peninsula, northwestern Western Australia. *Review of Palaeobotany and Palynology*, **120**, 17–39.

van der Kaars, S. *et al.* (2000). A late Quaternary palaeoecological record from the Banda Sea, Indonesia: patterns of vegetation, climate and biomass burning in Indonesia and northern Australia. *Palaeogeography, palaeoclimatology, Palaeoecology*, **155**, 135–53.

van Deusen, H.M. (1963). First New Guinea record of *Thylacinus*. *Journal of Mammalogy*, **44**, 279–80.

Van Huet, S. (1999). The taphonomy of the Lancefield swamp megafaunal accumulation, Lancefield, Victoria. *Records of the Western Australian Museum*, **supplement 57**, 331–40.

Van Huet, S. *et al.* (1998). Age of the Lancefield megafauna: a reappraisal. *Australian Archaeology*, **46**.

van Tets, G.F. (1985). *Progura gallinacea*: the Australian giant megapode, in *Kadimakara: extinct vertebrates of Australia* (eds P.V. Rich & G.F. van Tets), pp. 195–8. Pioneer Design Studio, Melbourne.

Vanderwal, R. & Fullagar, R. (1989). Engraved *Diprotodon* tooth from the Spring Creek locality, Victoria. *Archaeology in Oceania*, **24**, 13–16.

Veth, P. (1989). Islands in the interior: a model for the colonization of Australia's arid zone. *Archaeology in Oceania*, **24**, 81–92.

—— (1996). Current archaeological evidence from the Little and Great Sandy deserts, in *Archaeology of northern Australia* (eds P. Veth & P. Hiscock), pp. 50–65. Anthropology Museum, University of Queensland, Brisbane.

Wakefield, N.A. (1963). Mammal sub-fossils from near Portland, Victoria. *Victorian Naturalist*, **80**, 39–45.

—— (1967). Preliminary report on McEachern's Cave, S.W. Victoria. *Victorian Naturalist*, **84**, 363–83.

Wallace, A.R. (1876). *The Geographical Distribution of Animals, with a study of the relations of living and extinct faunas as elucidating past changes of the earth's surface*, vol. 1. Harper and Brothers, New York.

Wallis, L.A. (2001). Environmental history of northwest Australia based on phytolith analysis at Carpenter's Gap 1. *Quaternary International*, **83–85**, 103–17.

Walsh, G. (2000). *Bradshaw Art of the Kimberley*. Takarakka Nowan Kas Publications, Brisbane.

Walsh, G.L. & Morwood, M.J. (1999). Spear and spearthrower evolution in the Kimberley region, N.W. Australia: evidence from rock art. *Archaeology in Oceania*, **34**, 45–58.

Wang, X. *et al.* (1999). A record of fire, vegetation and climate through the last three glacial cycles from Lombok Ridge core G6-4, eastern Indian Ocean, Indonesia. *Palaeogeography, Palaeoclimatology, Palaeoecology*, **147**, 241–56.

Warneke, R.M. (1978). The status of the koala in Victoria, in *The koala* (ed. T.J. Bergin), pp. 109–14. Zoological Parks Board of New South Wales, Sydney.

Watts, C.H.S. & Aslin, H.J. (1981). *The Rodents of Australia*. Angus & Robertson, Sydney.

Wells, R.T., Horton, D.R. & Rogers, P. (1982). *Thylacoleo carnifex* Owen (Thylacoleonidae): marsupial carnivore?, in *Carnivorous Marsupials* (ed. M. Archer), vol. 2, pp. 573–86. Royal Zoological Society of New South Wales, Sydney.

Westerman, M., Springer, M.S., Dixon, J. & Krajewski, C. (1999). Molecular relationships of the extinct pig-footed bandicoot *Chaeropus ecaudatus* (Marsupialia: Perameloidea) using 12S rRNA sequences. *Journal of Mammalian Evolution*, **6**, 271–88.

White, J.P. & O'Connell, J.F. (1982). *A Prehistory of Australia, New Guinea and Sahul*. Academic Press, Sydney.

White, P. & Flannery, T. (1995). Late Pleistocene fauna at Spring Creek: a re-evaluation. *Australian Archaeology*, **40**, 13–16.

Wilkinson, C.S. (1885). President's address. *Proceedings of the Linnean Society of New South Wales*, **9**, 1207–41.

Williams, C.K. et al. (1995). *Managing Vertebrate Pests: rabbits*. Australian Government Publishing Service, Canberra.

Williams, M. *et al.* (2001). The enigma of a late Pleistocene wetland in the Flinders Ranges, South Australia. *Quaternary International*, **83–85**, 129–44.

Woinarski, J.C.Z., Milne, D.J. & Wanganeen, G. (2001). Changes in mammal populations in relatively intact landscapes of Kakadu National Park, Northern Territory, Australia. *Austral Ecology*, **26**, 360–70.

Woodburne, M.O. (2003). Monotremes as pretribosphenic mammals. *Journal of Mammalian Evolution*, **10**, 195–248.

Woodburne, M.O. & Case, J.A. (1996). Dispersal, vicariance, and the Late Cretaceous and Early Tertiary land mammal biogeography from South America to Australia. *Journal of Mammal Evolution*, **3**, 121–61.

Wright, K.L. (1997). An examination of the commensal interaction between the Australian native dung beetle, *Onthophagus peramelinus* and the rufous bettong *Aepyprymnus rufescens*. Hons thesis, James Cook University, Townsville.

Wright, R.V.S. (1986). New light on the extinction of the Australian megafauna. *Proceedings of the Linnean Society of New South Wales*, **109**, 1–9.

Wroe, S. (2002). A review of terrestrial mammalian and reptilian carnivore ecology in Australian fossil faunas, and factors influencing their diversity: the myth of reptilian domination and its broader ramifications. *Australian Journal of Zoology*, **50**, 1–24.

Wroe, S., Argot, S. & Dickman, C. (2004a). On the rarity of big fierce carnivores and primacy of isolation and area: tracking large mammalian carnivore density on two isolated continents. *Proceedings of the Royal Society of London B*.

Wroe, S., Crowther, M., Dortch, J., & Chong, J. (2004b). The size of the largest marsupial and why it matters. *Proceedings of the Royal Society of London B* (Suppl.), **271**, S34–S36.

Wroe, S. & Field, J. (2001a). Giant wombats and red herrings. *Australasian Science*, **22** (10), 18.

—— (2001b). Mystery of megafauna extinctions remains. *Australasian Science*, **Sept**, 21–5.

Wroe, S., Field, J., Fullagar, R. & Jermiin, L.S. (2004). Megafaunal extinction in the Late Quaternary and the global overkill hypothesis. *Alcheringa*, **28**, 291–330.

Wroe, S. & Johnson, C.W. (2003). Bring back the devil. *Nature Australia*, **271**, 1203–11.

Wroe, S., McHenry, C. & Thomason, J. (2005). Bite club: comparative bite force in big biting mammals and the prediction of predatory behaviour in fossil taxa. *Proceedings of the Royal Society of London B*, **272**, 619–25.

Wroe, S. *et al.* (2003). An alternative method for predicting body mass: the case of the Pleistocene marsupial lion. *Paleobiology*, **29**, 403–11.

Yibarbuk, D. *et al.* (2001). Fire ecology and Aboriginal land management in central Arnhem Land, northern Australia: a tradition of ecosystem management. *Journal of Biogeography*, **28**, 325–43.

Yokoyama, Y., Purcell, A., Lambeck, K. & Johnston, P. (2001). Shore-line reconstruction around Australia during the Last Glacial Maximum and Late Glacial Stage. *Quaternary International*, **83–85**, 9–18.

Yom-Tov, Y. (1985). The reproductive rate of Australian rodents. *Oecologia*, **66**, 250–5.

GLOSSARY

Note: The geological time scale is summarised, and the names of the intervals in the scale are shown, in the timeline on page x.

Araucariaceae The family of tall-growing native conifers that includes the bunya pine, hoop pine and Norfolk Island pine.

arboreal A mammal that spends its life in trees (e.g. koala, tree kangaroos).

artefact An object made or modified by people.

Asteraceae The daisy family.

browser A species that feeds on the leaves or woody tissue of shrubs and trees.

C3, C4 The two major photosynthetic pathways in plants. C3 plants are trees, shrubs and herbs. C4 plants are mostly grasses, especially from tropical or arid regions (although some arid-zone shrubs are also C4).

Casuarinaceae The family of casuarinas, or she-oaks.

chenopod A shrub of the family Chenopodiaceae, which includes the bluebushes and saltbushes common in the arid and semi-arid regions of Australia.

co-evolution (co-evolved) The process by which members of two or more species contribute reciprocally to the forces of natural selection that they exert on each other, resulting in the evolution of traits reflecting strong interspecific interactions (as between pairs of predator and prey species).

commensal A species that benefits from its association with a host species, while having no effect on that host species.

dimorphism The co-occurrence of two different structural forms in populations of a species.

ectotherm An animal whose body temperature is determined largely by the temperature of the environment, not by its own metabolism.

endotherm An animal whose body temperature is maintained largely by its own metabolic heat production.

ENSO El Niño/Southern Oscillation: the weather system, driven by a shifting gradient of air pressure and water temperature in the Pacific, that brings periodic multi-year droughts (El Niño events) to much of Australia.

Eutherian The mammalian subclass that includes the living placental mammals plus extinct related groups, to the exclusion of marsupials and their extinct relatives.

frugivore A species that feeds mainly on fruit.

Gondwana The name given to the southern continental landmasses (including India) when they were connected in the Cretaceous and earlier.

grazer A species that feeds predominantly on grass.

Greater Australia The continuous land area of New Guinea, Australia and Tasmania that emerges when sea levels are low under glacial climates.

Holocene The last ten thousand years of earth history.

insectivore An animal that feeds mainly on insects.

keystone species A species that provides ecological benefits to many other species in its habitat, to the extent that the abundance of those other species would decline substantially were it to disappear.

kyr Abbreviation of 'thousands of years'.

late Pleistocene The portion of the Pleistocene that takes in the last glacial cycle, i.e. 130 to 10 kyr ago.

LGM Abbreviation of 'last glacial maximum', i.e. the coldest period of the last glacial cycle, from about 35 to just after 20 kyr ago.

marsupials The 'pouched mammals', which characteristically give birth to very small young after a short gestation and which occur almost exclusively in Australia, New Guinea and South America. About half Australia's native land mammals are marsupials, the non-marsupials being monotremes, rodents, bats and the dingo.

mesopredator A small or medium-sized predator in an ecological community containing large, dominant predators; mesopredators are typically versatile and capable of living at high population densities.

midden A compacted pile of organic debris accumulated as a result of habitual feeding by people (or animals) in the same spot.

monotremes The mammal group represented by the living egg-laying mammals (the platypus and echidnas).

mycorrhiza Symbiotic association of plant roots and specialised fungi in which fungal hyphae envelop or penetrate the fine roots of the plant, and exchange soil nutrients and moisture from the soil for carbohydrates from the host plant.

Myrtaceae The family of sclerophyll plants that includes eucalypts, *Melaleuca* (paperbarks), *Leptospermum* (tea-trees) etc.

Nothofagus The genus of southern beech trees and shrubs, found mainly in southern Australia and southern South America (some species occur in warm and very wet climates).

palaeoecology The study of ecological conditions in the deep past, mainly using indicators from the fossil record.

phyllode A flattened leaf-stem that functions as a leaf. In many acacias, the 'leaves' are phyllodes.

placental mammal Mammals that reproduce via a long gestation, as distinct from the marsupials and monotremes. All living mammals other than marsupials and monotremes are placental mammals.

Pleistocene The geological epoch that lasted from about 2.6 million to 10 000 years ago.

Poaceae The grass family.

Prototheria One of the two major sub-divisions of the mammals (the other is the Theria), characterised by a primitive dentition; once diverse but represented among living species only by the monotremes.

Quaternary The geological period that includes the Pleistocene and the Holocene; that is, the interval from around 2.6 million years ago to now.

regression The statistical procedure for describing the effect of change in one variable on the value of another variable.

sclerophyll A leaf form typified by thickened, tough and sometimes small leaves; associated with long leaf lifespan, characteristic of many non-rainforest plants in Australia and an evolutionary response to moisture or nutrient shortage.

stratigraphy The relative positions of layers in a geological or sedimentary sequence.

sub-fossil Remains of an animal or plant that have been naturally preserved for a long period (e.g. in the dry conditions of a cave floor) but not yet fully buried by sediments.

terrestrial Living on the ground, as distinct from in trees.

Tertiary The geological period lasting from the end of the Cretaceous, 65 million years ago, to the beginning of the Pleistocene, about 2.6 million years ago.

wet sclerophyll forest Tall forest with a closed or nearly closed canopy of eucalypts, and a complex mid-storey and under-storey of moisture-loving plants.

ILLUSTRATIONS

PLATES

FIGURES

TABLES

INDEX

Note: Illustrations are shown with either bold page numbers or as Plates. Page numbers ending in 't' refer to tables.